VARIATIONAL PRINCIPLES
IN DYNAMICS
AND QUANTUM THEORY

Variational Principles in Dynamics and Quantum Theory

Third Edition

Wolfgang Yourgrau
Stanley Mandelstam

1968

W. B. SAUNDERS COMPANY

PHILADELPHIA

First published 1955
Revised and reprinted 1956
Second edition 1960
Reprinted 1962
Third edition 1968

PUBLISHERS:
SIR ISAAC PITMAN AND SONS LTD.
Pitman House, Parker Street, Kingsway, London, W.C.2

SOLE DISTRIBUTORS FOR THE WESTERN HEMISPHERE
W. B. SAUNDERS COMPANY – PHILADELPHIA

530.1
Y 82 V

Library of Congress Catalog Card Number 68—9169

MADE IN GREAT BRITAIN AT THE PITMAN PRESS, BATH

Dedicated to the Memory of
MAX PLANCK

Preface to the Third Edition

WE were not certain whether or not a new and considerably enlarged edition of this book was worth while presenting to the reader. The need, however, for a chapter treating variational principles in hydrodynamics was stressed by colleagues and students alike. Especially the late Erwin Schrödinger and Robert Oppenheimer, as well as Werner Heisenberg, advised us to add this (rather exacting) section to our monograph in order to satisfy an earnest demand.

Fortunately, Laurence Mittag (Yale) and Michael J. Stephen (Yale) generously offered to collaborate with one of us (W.Y.) on planning and writing this new Chapter 13. And thus we submit to the reader interested in variational principles in general, and in their relevance to hydrodynamics in particular, this third edition in the hope that he will find the whole subject as fascinating as we did—and still do.

WOLFGANG YOURGRAU
UNIVERSITY OF DENVER,
COLORADO.

STANLEY MANDELSTAM
UNIVERSITY OF CALIFORNIA,
BERKELEY.

Preface to the Second Edition

THE main alteration in this revised and enlarged edition consists of the incorporation of a new chapter on the Feynman and Schwinger principles in quantum mechanics. Further, a paper by one of us (W. Y.) and C. J. G. Raw on variational principles and chemical reactions, which appeared in *Il Nuovo Cimento*, is included as a separate appendix.

The authors wish specially to thank Professor E. Schrödinger for his invaluable criticisms and suggestions. Indeed, some of his comments proved to be so directly relevant that we have inserted them as postscripts to §§10 and 11. We are also indebted to Dr. A. J. van der Merwe for many helpful and stimulating discussions during the crucial stages of this edition.

The generous interest in the book expressed by Earl Russell, Prince L. de Broglie, Dr. J. R. Oppenheimer and Professors M. Born, M. von Laue and A. Speiser has greatly encouraged the authors in the preparation of this new edition.

Financial assistance for the research recorded in the present edition has been provided by the Bollingen Foundation and the South African Council for Scientific and Industrial Research; this support is gratefully acknowledged.

W. YOURGRAU

DONALD J. COWLING PROFESSOR OF
 PHILOSOPHY,
CARLETON COLLEGE,
MINNESOTA,
U.S.A.

FORMERLY VISITING RESEARCH PROFESSOR,
MINNESOTA CENTER FOR PHILOSOPHY OF
 SCIENCE,
UNIVERSITY OF MINNESOTA.

S. MANDELSTAM

ASSISTANT RESEARCH PHYSICIST,
UNIVERSITY OF CALIFORNIA, BERKELEY.

FORMERLY RESEARCH FELLOW,
DEPARTMENT OF PHYSICS,,
COLUMBIA UNIVERSITY, NEW YORK.

Preface to the First Edition

This study was prompted by the exceptional emphasis that Planck laid, both in his teaching and writing, upon the need for a methodical investigation into the scope and nature of the principle of least action. Our intention was to follow his suggestion of co-ordinating the historical, mathematico-physical and philosophic aspects of variational principles. It accordingly became necessary to select certain fields of the subject in order to concentrate upon those applications that are most relevant to modern physics.

In the sections devoted to dynamics we were faced with the alternatives of discussing only variational principles proper or of intensively scrutinizing the analytical theory. However, a brief review of Lagrange's and Hamilton's equations and of contact transformations commended itself in the endeavour to find a mean between these two possibilities. On the other hand, we have refrained from treating the principle of virtual displacements and d'Alembert's principle, and have simply considered the constraints in dynamical systems to be caused by internal forces. For the same reason, such topics as the Poisson bracket and the Legendre transformation have not been examined.

As there is a close connexion between variational principles and quantum theory, we have thought it appropriate to incorporate both subjects in one treatise, which may be of interest to students seeking insight into theoretical developments that led to quantum mechanics, without consulting more elaborate works. Similar considerations rendered it desirable to allude to the derivation of Schrödinger's wave equation, which employs concepts analyzed in the earlier part of this monograph.

The reader will notice that, to indicate the manner in which the theories mentioned were evolved, special attention has been directed to the outstanding contributions made by the respective investigators. Considerable difficulty was encountered in obtaining certain original papers and we are most indebted to Professor A. Speiser, University of Basle, for his kindness in providing us with the proofs of Vol. 24 of Euler's *Opera omnia*, edited by C. Carathéodory. We wish to acknowledge also the assistance received from the Library of the University of Heidelberg, the Bibliothèque Nationale, Paris, and the South African Public Library, Cape Town.

A grant, allocated by the Research Committee of the University of the Witwatersrand, has enabled us to prepare the manuscript for publication. In particular are we most obliged to Professor J. M. Hyslop, Department of Mathematics, and Professor D. J. O'Connor, Department of Philosophy, both of the University of the Witwatersrand, for instructive comments and criticism. Our grateful thanks are due to Mr. N. Garber, F.R.C.S., Mr. M. G. Clarke, Miss P. Gresty, Mrs. M. Kalk, Miss A. Callinicos, Mrs. I. Elsas and Mr. A. S. Mashugane, for substantial aid in completing this work. And we would also like to say how much we owe to the editorial staff of Sir Isaac Pitman & Sons Ltd., for their untiring co-operation and their readiness to meet all our wishes, and to the technical staff for their excellent printing.

Finally, our deepest gratitude to Thella Yourgrau for her inestimable encouragement and faith.

UNIVERSITY OF WITWATERSRAND
JOHANNESBURG.

W. YOURGRAU,
S. MANDELSTAM.

September, 1952

Contents

§ 1

Prolegomena

FROM the earliest times philosophers and scientists have tended to reduce the manifold phenomena of nature to a minimum of laws and principles. Thus, the adherents of the Milesian school (*c.* 600–520 B.C.), the so-called Ionian physiologists, postulated a single substrate from which all substances comprising the cosmos were derived. In contradistinction to the physical, materialistic systems of these early Greek thinkers, Pythagoras (*c.* 530 B.C.) and his followers regarded number as the prime entity governing the universe.

The foundation of Pythagoras' cosmology was the *tetractys of the decad*, the progression of the integers 1,2,3,4, whose sum is equal to the "perfect" number ten. This integer is the fourth of the "triangular numbers" $1,3,6,10,15, \ldots, \frac{1}{2}n(n + 1), \ldots$ which played a dominant rôle in Pythagorean doctrine. Nature in its entirety, according to this doctrine, was composed of various *tetractyes*, such as the geometrical ascendency of point, line, surface, solid or the primordial elements earth, water, air, fire. The property of "Fourness" was evident also in the Pythagorean *quadrivium*: geometry, arithmetic, astronomy and music.

By an amazing experimental achievement, Pythagoras revealed the fundamental laws of acoustics and observed that consonant musical intervals are expressed by simple ratios,—by fractions whose numerator and denominator are members of the *tetractys* 1,2,3,4. The notion of ἁρμονία pervades the whole philosophy of Pythagoreanism, but the term "harmony" at that time meant tuning or scale or octave. Many interpretations have been read into the phrase "harmony of the spheres," attributed to the founder of the order. It is probable that

Pythagoras conceived the intervals of the fourth, the fifth and the octave to correspond with the three rings of Anaximander, who had imagined the sun, the moon and the stars to be transported on three wheels around the earth. As the heavenly bodies revolved, they produced a musical harmony of transcendent beauty to which we, however, had grown deaf. The concepts of harmony and number are thus inextricably interwoven in the Pythagorean cosmology, and acoustical laws appeared to embrace the whole structure of the universe.

The extent to which Pythagoras appreciated the discovery of the relations between consonances and numerical ratios is summed up in his maxim that all things are numbers. This dictum he applied, *inter alia*, to his theory of the planets. The celestial bodies—some authorities hold—as well as the earth were considered to be spheres and to move in circles, because circle and sphere were deemed to be "perfect" figures. Apart from the fixed stars, seven heavenly bodies were visible; to these were added the earth and the central fire, Hestia (not identical with the sun), about which the whole system turned. But, since ten was the perfect number containing all natural things, the Pythagoreans postulated a tenth body, the "Counter-earth" which was supposed to balance the revolution of the earth around the central fire. Both Hestia and the Counter-earth were unobservable to the human eye.

Despite the unquestionable mysticism prevailing throughout his cosmological and physical speculations, Pythagoras made momentous and lasting contributions to mathematical knowledge. In particular, he introduced the notions of axiom and proof into geometry: the very terms mathematics, theory and philosophy, as they are known today, were originated by the Pythagoreans.

It is beyond doubt that the essence of Pythagoreanism lies in the explication of phenomena by a philosophy of nature simultaneously *a priori* mathematical and mystical. Further, is it not remarkable that Pythagoras was enamoured of the preponderance of whole numbers, an idea which, during the first quarter of this century, reappeared in quantum theory? Mathematics and physics were in his time not distinguished from one another, and, in contrast to Plato, the method of induction was sanctioned and utilized, i.e., the Pythagoreans resorted to the realm of sensory experience. In addition to these considerations it is safe to maintain that theology, too, was closely linked to mathematics,—a combination again to be encountered in the work of Maupertuis. And permeating the general outlook of Pythagorean teaching seems to have been the tendency to arrive, through few premises, at conclusions which only an experimentalist would venture to draw, because he alone could verify his findings.

The emphasis upon mathematical reasoning and abstract thought in general, which the genius of Pythagoras had instigated, reached its climax in the theoretical acumen of Plato (428/7–348/7 B.C.), who rejected the experimental method with ardour and contempt. In his view, no precise study of the ever-changing phenomena in the natural universe was possible, and it was only in the philosophic theory of forms and in the science of pure mathematics that absolute knowledge could be attained. These contemplative disciplines of the intellect dealt with objects timeless and invariant, known independently of experience and existing logically prior to the material world, which could at most be merely an approximation to eternal forms or ideas.

In the *Timaeus*, Plato relates how the Demiurge, the "architect of the world," created the cosmos by reducing the primal state of chaos to an ordered pattern.

> . . . he desired that all things should come as near as possible to being like himself. That this is the supremely valid principle of becoming and of the order of the world, we shall most surely be right to accept from men of understanding. Desiring, then, that all things should be good and, so far as might be, nothing imperfect, the god took over all that is visible—not at rest, but in discordant and unordered motion—and brought it from disorder into order, since he judged that order was in every way the better . . . . the fitting shape would be the figure that comprehends in itself all the figures there are; accordingly, he turned its shape rounded and spherical, equidistant every way from centre to extremity—a figure the most perfect and uniform of all; for he judged uniformity to be immeasurably better than its opposite.[1]

We can infer from his dialogues that Plato, himself a mathematician of no mean power, continued the Pythagorean legacy that number rules the universe. This can be seen, for instance, in his reference to the *tetractyes* consisting of the geometrical progressions 1,2,4,8 and 1,3,9,27, that include those ratios of which the whole cosmos is constructed. "God ever geometrizes,"—θεὸς ἀεὶ γεωμετρεῖ. However spurious this quotation may be, it nevertheless characterizes Plato's "vision of truth" most tellingly.

The modern student will in vain search the writings of Plato for the concept of exact laws of nature as it is employed by us. In truth, the cosmology of the *Timaeus* is a myth of creation, a cosmogony rather than an astronomical treatise. Yet it is still significant for our purpose, in that it provides the impress of a great thinker upon a conceptualized representation of the phenomenal world through such ideas as simplicity, uniformity, order and perfection.

It is in Aristotle (384–322 B.C.), however, that a more definite formulation of a simplicity hypothesis is initially recorded. All motion, according to Aristotle, is either rectilinear, circular, or a combination of the two, because these are the only "simple movements." Upward motion is motion towards the centre, downward motion leads away from it, while circular motion is movement around the centre. Further, all motion is described as being either *natural* or *unnatural* to the moving body; upward movement is natural to fire, downward movement natural to earth. Since circular movement is "perfect," i.e., eternal and continuous, it must be natural to some system, and moreover to some system which is "simple" and "primary." Aristotle therefore infers that the celestial bodies revolve in circles.

In another passage of *De caelo*, Aristotle mentions, as an explanation for the circularity of planetary motion, the fact that of all curves enclosing a given area the circle possesses the shortest perimeter. "Again," he wrote,

> if the motion of the heaven is the measure of all movements whatever in virtue of being alone continuous and regular and eternal, and if, in each kind, the measure is the minimum, and the minimum movement is the swiftest, then, clearly, the movement of the heaven must be the swiftest of all movements. Now of lines which return upon themselves the line which bounds the circle is the shortest; and that movement is the swiftest which follows the shortest line. Therefore, if the heaven moves in a circle and moves more swiftly than anything else, it must necessarily be spherical.[2]

These remarks are of some concern to us, as the transition from the belief in simplicity to a "minimum" postulate is here explicitly carried out for the first time.

Aristotle's conception of nature stood in marked contrast to Pythagoreanism and Platonism in that it grossly underestimated the importance of mathematics. Logic, on the other hand, was overrated—an ill fortune, for the intrinsic limitations of traditional formal logic rendered it completely inadequate.

Although we must not read too much into Aristotle's reflections, they nevertheless indicate a less mystical attempt at interpreting the data of a specific science, namely astronomy, after a certain aesthetic and metaphysical ideal of simplicity. The state of scientific knowledge in that era was in no way ripe for such an undertaking and Aristotle was compelled—as von Laue remarked—"to include, in his otherwise grandiose system of natural science, only a few concepts, taken rather non-critically from superficial observations, and their logical or oftentimes merely sophistical analysis." Such an attitude, if employed indiscriminately and applied *a priori*, may lead to sterility and therefore hamper empirical development.

Aristotle's minimum hypothesis, which occupied only a subordinate and scarcely noticeable position in his work, was clearly not dictated by any appeal to quantitative measurement and was not subject to rigorous scrutiny. Hero of Alexandria (*c.* 125 B.C.), however, proved in his *Catoptrics* a genuinely scientific minimum principle of physics. He showed that *when a ray of light is reflected by a mirror, the path actually taken from the object to the observer's eye is shorter than any other possible path so reflected.* This proposition was obtained from a generalization of the observed fact that in general, when light travels from one point to another, its path is a straight line, that is, it takes the shortest distance between these points. Owing to the historical significance of Hero's discovery for the subject of this treatise, his method of reasoning may be worth quoting.

> Practically all who have written of dioptrics and of optics have been in doubt as to why rays proceeding from our eyes are reflected by mirrors and why the reflections are at equal angles. Now the proposition that our sight is directed in straight lines proceeding from the organ of vision may be substantiated as follows. For whatever moves with unchanging velocity moves in a straight line. . . . For because of the impelling force the object in motion strives to move over the shortest possible distance, since it has not the time for slower motion, that is, for motion over a longer trajectory. The impelling force does not permit such retardation. And so, by reason of its speed, the object tends to move over the shortest path. But the shortest of all lines having the same end points is the straight line. . . . Now by the same reasoning, that is, by a consideration of the speed of the incidence and the reflection, we shall prove that these rays are reflected at equal angles in the case of plane and spherical mirrors. For our proof must again make use of minimum lines.[3]

Although Hero differed from Aristotle by demonstrating mathematically that his principle was in agreement with experimental data, he considered this principle to provide an 'explanation' of these data. His approach was therefore akin to Aristotle's in that he deduced his results from preconceived suppositions, and a certain similarity in outlook can be perceived between Aristotle's notion of simplicity and Hero's minimum condition. It was not, however, until the advent of Fermat in the 17th century that further attention was directed toward minimal principles of this kind.

Returning now to the simplicity postulate, we find that perhaps its most consistent advocate before the days of Einstein was William of Ockham (*c.* 1300–1347). The *doctor invincibilis*, one of the profoundest speculative minds of the scholastic period, is at present remembered mainly for his celebrated "razor": *Entia non sunt multiplicanda praeter necessitatem.*   In this form, the maxim is actually not to be found in his writings, but it has become intimately associated with his name, as its essence pervades the whole of his work.   *"Frustra fit per plura quod potest fieri per pauciora"*—it is futile to employ many principles when it is possible to employ fewer.   Ockham thus maintained that the number of hypotheses should not exceed the minimum necessary for explanation of the facts.

Any attempt at assessing Ockham's logico-metaphysical system will confirm that the trend of science and natural philosophy prevailing today has been foreshadowed by his outlook if not inaugurated by it.   In accordance with the scholastic legacy of St. Augustine and St. Thomas Aquinas, the theory of knowledge in all its diverse aspects was completely dependent on theology and ontology.   Science in the modern sense of the word was practically unknown, and what did exist was totally subservient to the Church.   It is without doubt the inalienable merit of this Franciscan schoolman that he sundered anew the calamitous union of faith and reason.   Followed to its conclusion, his secular attitude would adumbrate our present predisposition towards observation, experiment and theory.

The "razor" clearly entails the rejection of the Platonic conception that universals exist apart from and prior to so-called real things.   Ockham was, indeed, probably the foremost nominalist of the middle ages.   He emphasized the distinction between statements referring to language and statements referring to things.   Denying independent reality to universals, i.e., mere abstractions, he anticipates the phenomenalist and positivist schools and Mach's economy of thought.   It should, however, be fully understood that, despite his professed nominalism, his name can never be allied with movements such as empiricism, logical positivism, or pragmatism; for it is obvious that Ockham could not have reckoned with the meaning and implications of such contemporary theories.

Ockham's principle, while indicating a viewpoint similar to the simplicity hypothesis of Aristotle, should none the less be distinguished from it.   The Greek thinker held that nature possesses an immanent tendency to simplicity, whereas Ockham demanded that in *describing* nature one should avoid unnecessary complications. Examples of both doctrines occur frequently in the history of science; Copernicus (1473–1543), for instance, reiterated his belief in the simplicity prevalent *in rebus naturae*. Even he, hailed as the harbinger of a new era in natural science, displayed strong Pythagorean features. Perhaps echoing the cosmology of the *Timaeus*, then common property among western philosophers, his treatise abounds in references to Platonic ideas. He regarded the cosmos as spherical, a "divine body" endowed with the perfection of its creator, and rejoiced in the regularity and order of the world. Circular motion was supposed to be proper to all complete objects. Like the Greeks, Copernicus appraised the state of rest as more noble than that of motion. Yet he also boasted that his own heliocentric theory contained fewer and simpler hypotheses than the geocentric model of Ptolemy. We thus encounter Pythagorean-Platonic mysticism and scientific reasoning in a combination which, from a modern standpoint, must appear to us as strange.

With his resolution to submit every physical and astronomical law to the test of experiment and observation, Kepler (1571–1630) contributed largely to the inauguration of the present scientific age. It is therefore surprising to perceive in his work copious signs of superstition and a keen devotion to astrology. Neo-Platonic and religious conceptions are even more evident than in Copernicus. So was he, still under the spell of apriorism, anxious to interpret the universe as motivated by a mathematico-aesthetic numerical harmony and exhibiting a surpassing simplicity and unity—*natura simplicitatem amat*. Kepler dissented from the Aristotelian metaphysics of his day and maintained that the Copernican system was not merely a convenient hypothesis, but was a true image of nature for the very reason that it was mathematical in form and amenable to quantitative measurement. However, rather than attempt to force data into an artificial agreement with any arbitrary rational system, he differed from the Pythagoreans and moulded his theory to suit the facts. He did indeed take us part of the way from the Pythagorean to the modern attitude respecting the function of mathematics in science, and considered his three laws as noteworthy in that they demonstrated a mathematical relation between phenomena that had previously been unconnected.

By far nearer to contemporary rational thinking is the "new science" of Galileo (1564–1642) who repudiated apodictically the mystical indulgence which assumed so prominent a position with Kepler. True, Galileo did found the method of analysis and induction in physics, and discarding the view, so typical of Greek cosmology, that nature is an organism, he espoused a decidedly mechanistic interpretation. On closer examination, however, there is again ample evidence of his interest in Pythagorean and Neo-Platonic doctrines, which had been disseminated during the mediaeval period and the early renaissance. As a pithy illustration, we need only cite the oft-quoted passage: "Philosophy is written in that great book which ever lies before our eyes—I mean the universe—but we cannot understand it if we do not first learn the language and grasp the symbols, in which it is written. This book is written in the mathematical language, and the letters are triangles, circles and other geometrical figures, without which means it is humanly impossible to comprehend a single word."[4] His work is premised by the deeply rooted conviction of a simple, ordered world, free from arbitrariness and disclosing a geometrical regularity. Thus, after he had approximately verified by observation that a falling body suffered equal increments of velocity in equal times, he thought it unnecessary to have recourse to further and more accurate experiments. Galileo conceived of this law as the most natural and fitting; it required confirmation merely by consulting nature.

These brief remarks serve to indicate that Galileo, like Kepler, commenced his investigations with aprioristic presuppositions, though both insisted that the results of abstract thinking should, as a criterion for their validity, correspond to the findings of empirical research.

We have dwelt at some length upon the welter of ideas which adorned the intellectual background of those centuries. So significant were they for the beginning of modern scientific philosophy that they penetrated even the *Principia* of Newton (1642–1727). In the *Regulae philosophandi* we meet, commingled, both the demand for a minimum of hypotheses in the treatment of natural phenomena and the truism, by then well-worn, that simplicity is a characteristic pertaining to nature itself. *"We are to admit no more causes of natural things than such as are both true and sufficient to explain their appearances.* To this purpose the philosophers say that Nature does nothing in vain, and more is in vain when less will serve; for Nature is pleased with simplicity, and affects not the pomp of superfluous causes."[5] It was through a concrete application of such postulates that "minimum" principles first reappeared in physics since the time of Hero, whilst similar ideas provided the conceptual basis for the later formulation of these minimum postulates. Hints in this direction are manifest, for instance, in Leonardo da Vinci's comment that when falling in a straight line a body takes the shortest path, and in the adage, current in this period, that "*natura semper agit per vias brevissimas.*"

With the philosophy of Leibniz (1646–1716), however, we approach still more closely to the explicit recognition of minimum conditions in nature. He was convinced that the Supreme Being does not create anything remotely resembling disorder; one cannot, in fact, even conceive of some haphazard occurrence in the universe. More strongly than any other Rationalist either before him or since, Leibniz propagated the doctrine of simplicity prevailing throughout the real world. This idea is proximately derived from the *Théodicée* and the *Discours de métaphysique* where he attempts to prove with logical cogency that this world of ours is the best of all possible worlds, exhibiting "the greatest simplicity in its premises and the greatest wealth in its phenomena." Conforming to this metaphysical tenet, the apostle of a *Scientia generalis* advocated an analogous simplicity in our interpretation of nature, and therefore demanded, like Ockham, that we should abhor unnecessary multiplicity of suppositions. Hence the reason for the adoption of the Copernican theory in astronomy—it is simpler, not truer, than the geocentric theory. In this outlook Leibniz anticipates the positivism of Mach. The way in which such "economy" postulates act as a foundation for the development of minimum considerations is distinctly revealed in Leibniz's contention that "the perfectly acting being . . . can be compared to a clever engineer who obtains his effect in the simplest manner one can choose."

It is interesting also to note that the French philosopher Malebranche (1638–1715), in his *Recherche de la vérité*, arrived at a similar view which he called the "Economy of Nature." Again, as might be expected, his method of approach was purely speculative, in agreement with the general philosophy of the Cartesian school.

It is obvious, if we reflect on these facts, that some of the greatest minds in the history of human thought had already affirmed a fervid though scientifically unfounded belief in a rational, regularly functioning universe and had thereby implied the anthropomorphic conception of orderliness and simplicity in nature proper. A similar anthropomorphism is exemplified in notions like force, *vis viva*, work, energy, power, etc., which have only gradually lost their human connotation. Does the idea of bodies being moved by such forces differ much from that of bodies moving so as to involve the minimum expenditure of some quantity?

But the fallacious though venerable belief in an economic and simple pattern conspicuous in the material world did not remain uncriticized. Francis Bacon (1561–1626) castigated those Pythagorean features inherent in human nature, traits which impose our emotions and wishes upon the manner wherein we co-ordinate and integrate empirical facts. "And the human understanding," he wrote,

> is like a false mirror, which, receiving rays irregularly, distorts and discolours the nature of things by mingling its own nature with it. . . . The human understanding is of its own nature prone to suppose the existence of more order and regularity in the world than it finds. And though there be many things in nature which are singular and unmatched, yet it devises for them parallels and conjugates and relatives which do not exist. Hence the fiction that all celestial bodies move in perfect circles . . . . whence proceed sciences which may be called "sciences as one would."[6]

Though we may not agree with the zeal and self-righteousness of Bacon's attacks, yet they have incontestably laid bare the innate weaknesses of such aprioristic systems in science.

In conformity with the scope of our subject, the speculative facets of the thinkers under review have been emphasized. Historically by far more consequential were their positive contributions to natural science, contributions which transferred the emphasis from *a priori* reasoning to theories based upon observation and experiment. Hence, while the future exponents of least principles may have been guided in their metaphysical outlook† by the idealistic background we have described, they had nevertheless to present their formulations in such fashion that the data of experience would thus be explained. A systematic scrutiny of the individual chronological stages in the evolution of minimum principles can furnish us with profound insight into the continuous transformation of a metaphysical canon to an exact natural law.

---

† By "metaphysical outlook" we comprehend nothing but those general assumptions which are tacitly accepted by the scientist.

§ 2

Fermat's Principle of Least Time

It is in optics that we encounter minimum principles for the first time. The French mathematician Fermat (1601–1665) postulated that, no matter to what kind of reflection or refraction a ray is subjected, it travels from one point to another in such a way as to make the time taken a minimum. We have already seen how Hero of Alexandria had discovered a similar law for the particular case where the ray of light is reflected by a mirror. In view of the close relation between the two theorems, Hero's demonstration that his principle is equivalent to the customary law of reflection will now be cited.[3]

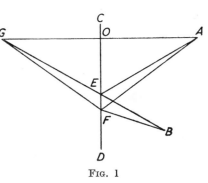

Fig. 1

In Fig. 1, A represents the source of light, B the eye, CD the mirror. AEB is the path actually taken, AFB any other possible path undergoing reflection at the mirror. Let AO be perpendicular to CD, and produce BE to meet AO produced in G. From the equivalence of the angles of incidence and reflection, it is immediately evident that $AE = EG$ and $AF = FG$; from the triangular inequality $GF + FB > GB$, it follows that

$$\text{path } AFB > \text{path } AEB. \qquad . \qquad . \qquad . \qquad (1)$$

But AFB was any possible path. Hence, of all paths undergoing reflection at the mirror, AEB is the shortest.

Hero then applies the same mode of reasoning to spherical mirrors (concave and convex).

On reflection it seems that Fermat, like Hero, arrived at his proposal intuitively rather than by considering observed facts. However, by applying the condition that the time taken along an adjacent path had to differ from the time actually taken by a quantity of second order, he did show that Snell's law of refraction conformed to his hypothesis.[7]

AB is the refracting surface separating the rarer medium above it from the denser medium below it; C is the light source.

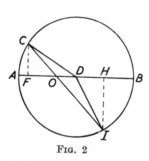

Fig. 2

Fermat constructs the circle $ACBI$ to meet the refracted ray at I. Let the "resistances"† of the rarer and denser media be proportional to m and b respectively. We have to postulate that m is less than b, "because the resistance of the rarer medium ought to be less than that of the denser medium, by an axiom which it is natural to adopt." For convenience, let b equal DF. Draw IH and CF perpendicular to AB. Denote DH by a, and the radius of the circle by n. Then, if we consider any adjacent path COI and e represents the length of DO,

$$CO^2 = n^2 + e^2 - 2be$$

and

$$OI^2 = n^2 + e^2 + 2ae.$$

Now the time taken along the path CDI is equal to $mn + bn$, and that taken along path COI is $m \cdot CO + b \cdot IO$. Hence

$$\sqrt{m^2n^2 + m^2e^2 - 2m^2be} + \sqrt{b^2n^2 + b^2e^2 + 2b^2ae} = mn + bn, \quad (2)$$

to quantities of the first order in e. Squaring, we obtain

$$m^2n^2 + m^2e^2 - 2m^2be + b^2n^2 + b^2e^2 + 2b^2ae$$
$$+ 2\sqrt{m^2n^2 + m^2e^2 - 2m^2be} \; \sqrt{b^2n^2 + b^2e^2 + 2b^2ae}$$
$$= m^2n^2 + 2n^2bm + b^2n^2,$$

or $2\sqrt{m^2n^2 + m^2e^2 - 2m^2be} \; \sqrt{b^2n^2 + b^2e^2 + 2b^2ae}$
$$= 2m^2be - 2b^2ae + 2n^2bm - m^2e^2 - b^2e^2.$$

† By "resistance" Fermat evidently refers to the reciprocal of the velocity.

Squaring again, it follows that

$$4(m^2b^2n^4 + 2m^2n^2b^2e^2 + 2m^2n^2b^2ae + m^2b^2e^4 + 2m^2b^2ae^3$$
$$- 2m^2n^2b^3e - 2m^2b^3e^3 - 4m^2b^3ae^2)$$
$$= 4m^4b^2e^2 + 4b^4a^2e^2 + 4n^4b^2m^2 + m^4e^4 + b^4e^4 - 8m^2b^3ae^2$$
$$+ 8m^3n^2b^2e - 4m^4be^3 - 2m^2b^3e^3 - 8b^3n^2mae + 4b^2m^2ae^3$$
$$+ 4b^4ae^3 - 4n^2bm^3e^2 - 4n^2b^3me^2 + 2m^2b^2e^4.$$

Hence, cancelling the terms not containing e (which are the same on both sides), and neglecting those of second and higher order in e, the equation reduces to

$$4(2m^2n^2b^2ae - 2m^2n^2b^3e) = 8m^3n^2b^2e - 8b^3n^2aem,$$

or
$$ma - mb = m^2 - ba.$$

Thus
$$(m - a)(m + b) = 0$$

and therefore
$$m = a, \qquad \qquad (3)$$

the negative root being clearly inadmissible. But a/FD is obviously equal to the ratio of the sine of the angle of incidence to that of the angle of refraction. Thus this ratio is constant and equal to the ratio of the velocities in the two media.

Although Fermat had hereby proved that Snell's law could be deduced from his principle of least time, this assumption nevertheless implied more than had been observed experimentally, since it was not then known that the refractive index was equal to the ratio of the velocities of light in the two media. In fact, Fermat used the result to corroborate his opinion (namely, that light travels more slowly in the denser medium) against the opposite view of Descartes. It seems evident, therefore, that the principle of least time was, as we have been led to expect, derived in part at least from *a priori* reasoning, and not by inductive argument.

Reverting to Fermat's method of demonstration, we must emphasize again the salient point that he merely restricts himself to proving that the time required for a light ray to traverse a neighbouring virtual path differs from the time actually taken by a quantity of the second order. In other words, *the variation between the time taken to travel along the actual path and that needed to cover an adjacent virtual path is zero.* This condition is necessary but not sufficient for the time to be a minimum. In this simple case, as Fermat showed elsewhere, and as will be seen below, the minimum condition can be proved;

but in more complicated cases the two conditions, viz., the minimal and the variational, are not equivalent. When the variation is zero, we may not even have an extreme value.

The first real justification of Fermat's principle was given by Huygens (1629–1695) who deduced, on the basis of the wave theory, that the refractive index between two media is equal to the ratio of the velocities of light in the media.[8] He thereby confirmed that Fermat's principle holds if the validity of wave

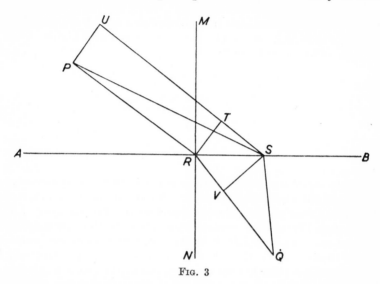

Fig. 3

optics is assumed. Huygens' proof may be worth quoting, because it demonstrated that the time taken by light to pass between two points is a real *minimum*; it is moreover much shorter than Fermat's proof of the same fact.

Let AB, Fig. 3, denote the refracting surface, P the starting point, Q the end point. Suppose PRQ is the actual path taken, PSQ any other path. Draw SU parallel to RP and PU perpendicular to RP, the two lines meeting in U. Let RT be a perpendicular on to SU and SV a perpendicular on to RQ. The velocities of the light in the media above and below AB are represented by v_1 and v_2 respectively. Huygens' work allows Snell's law to be written in the form

$$\frac{\sin i}{\sin r} = \frac{v_1}{v_2}, \qquad \cdot \quad \cdot \quad \cdot \quad \cdot \quad \cdot \quad (4)$$

where i is the angle of incidence PRM and r the angle of refraction NRQ.   From this it is easily seen that

$$\frac{TS}{RV} = \frac{v_1}{v_2}. \qquad . \quad . \quad . \quad . \quad (5)$$

Thus, the times required to traverse the distances TS and RV are equal.   Further, the times required to traverse the distances TU and RP are equal (since $TU = RP$), and the time required to traverse SQ is greater than that required to traverse QV (since $SQ > QV$).   Hence the time taken for light to travel along USQ is greater than that taken along PRQ and, as

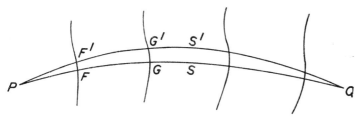

Fig. 4

$PS > US$, the time necessary to traverse PSQ is greater than that required to traverse PRQ.

A similar result can be proved if S is to the left of R.   It follows that, since S is any point on AB, the time taken to travel along PRQ is less than that taken to travel along any other path from P to Q.

The proofs so far developed are based upon ray optics, Huygens having merely referred to wave optics in order to obtain the ratio of the velocities of light in the different media.   This method, however, fails to elucidate the essential significance of the principle under consideration.   It seems beyond doubt that the derivation of Fermat's principle is best approached through direct application of the wave theory.   At the same time, such a demonstration is far more general than the preceding deductions which deal with simple cases only and need elaboration to be applicable to more complicated cases.

In Fig. 4, we depict the surfaces of constant phase of light waves in a heterogeneous medium, the orthogonal trajectories of these "wave surfaces" denoting the rays. PSQ is such a ray, and $PS'Q$ any adjacent path from P to Q.   Let PSQ cut the wave surfaces in F, G, H, $\ldots$, and let $PS'Q$ cut them in $F',G',H', \ldots.$ The time taken for the light to pass from one wave surface to another along a ray, being the time needed for the one wave surface to move to the position of the other, is independent of the particular ray chosen. The difference between the time required to pass from F to G and that required to pass from F' to G' depends, therefore, only upon the sine of the angle between $F'G'$ and the wave surfaces. Since this angle differs from $\pi/2$ by a small quantity of the first order, the sine differs from unity by a small quantity of the second order. The time taken to travel from F' to G' thus differs from that required to travel from F to G by a small quantity of the second order. Summing over the whole of PQ, we find that the variation between the time needed to traverse PSQ and the time needed to traverse a neighbouring path is zero (neglecting infinitesimals of the second order in the distance between the two paths).

At first sight, it may appear as if the time needed to traverse PSQ is *less* than that required to traverse any neighbouring path, because the distance along the neighbouring path between two wave surfaces cannot be less than the perpendicular distance between the two surfaces. But closer scrutiny reveals a factor of which we have as yet not taken account. It has been assumed that the wave surfaces are simple curves. Now it can admittedly be seen that, starting with the given wave surface through P, we can construct successive surfaces to the right of P for a certain distance, but it is by no means true that we can in general continue in this way for any arbitrarily great distance. It may occur that a ray adjacent to PQ cuts PQ at, say, a point R. The minimum condition in question holds therefore only for sufficiently small distances. The variational condition (but not the minimum condition) can easily be extended to arbitrary distances.

In Fig. 5, the lines AB, CD, EF, GH, ... are such that the distances between any line and the next but one are small enough to comply with the requirements of the preceding proof. Then the variation between the time taken for the actual path $PIKLM \ldots Q$ and for the virtual path $PIXLZ \ldots Q$ is zero, for the variation between IKL and IXL is zero, and we can divide up the whole path in a like manner. Similarly, the variation between path $PIXLZ \ldots Q$ and path $PNXYZ \ldots Q$ is zero. Combining these two results, it follows that the variation between the times taken in travelling along path $PIKLM \ldots Q$ and along path $PNXYZ \ldots Q$ is zero.

This proof of Fermat's principle on the basis of wave optics has been worked out here for heterogeneous media, but it is plainly valid also if there are in addition reflecting or refracting surfaces.

At this stage, we can conclude that Fermat's original assumption was not quite correct, since in the general case the variational condition has to be substituted for the more restricted minimal condition. In its variational form, the principle has lost the aesthetic fascination that the metaphysical postulate of *minima* intrinsic in nature appears to have possessed for mathematicians and philosophers alike in the time of Fermat. This change of aspect is important for our later discussion, as some physicists still profess to view "least" principles as a

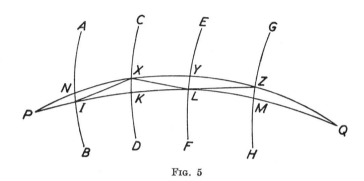

Fig. 5

confirmation of an *a priori* explication of a great many physical phenomena.   The conclusions at which Fermat arrived by such a method had to be modified to fit the experimental data.   His method of approach and the arguments which he implicitly adduced in favour of his fundamental law are, nevertheless, of more than historical interest, as they led eventually to a correct and experimentally verified result.

For future applications, it is most convenient to state Fermat's principle mathematically as

$$\delta \int_P^Q \frac{ds}{v} = 0, \quad . \quad . \quad . \quad . \quad (6)$$

where P and Q are the starting- and end-points of the path, v the velocity at any point and ds an element of the path.   The equation indicates that the variation of the integral is zero, i.e., the difference between this integral taken along the actual path and that taken along a neighbouring path is an infinitesimal quantity of the second order in the distance between the paths.

It must still be specified whether v in eq. (6) denotes the phase or the group velocity; they are not the same in a dispersive medium. An examination of the general (wave optics) proof shows immediately that v represents the phase velocity, and we accordingly adopt this interpretation. Fermat's idea must therefore be further modified; translated into the language of wave optics, the original formulation would entail that the time required for a *group* of waves to pass between two points should be a minimum, and that v would thus be the group velocity.

§3

The Principle of Least Action of Maupertuis

THERE was enunciated in 1744 a scientific dictum that was destined to gain greatest prominence in natural philosophy and that has perhaps been the subject of more controversy than any similar proposition.[9] The French mathematician Maupertuis (1698–1759) announced in that year *le principe de la moindre quantité d'action*, the famous principle of least action.† In the spirit of the Platonic-Pythagorean cosmology and in close conformity with Leibniz, he decided that *"la Nature, dans la production de ses effets, agit toujours par les moyens les plus simples."* Following in the footsteps of Hero and Fermat, he then proclaimed that this simplicity causes nature to act in such a way as to render a certain quantity, which he termed the "action," a minimum. To quote his own words: *"Lorsqu'il arrive quelque changement dans la Nature, la quantité d'action, nécessaire pour ce changement, est la plus petite qu'il soit possible."* Almost overbearingly did he postulate that the action must depend on the mass, the velocity and the distance; he therefore defined action as the product of these three factors.

Maupertuis' definition of action was very obscure, owing to the fact that the distance covered by a moving body varies with time, and he failed to specify the time interval for which the product *m.v.s* is to be computed. In each of his examples he assigned a different meaning to the action, so that every case yielded, *mutatis mutandis*, the desired result.

It was while trying to discover a rational and metaphysical basis for geometrical optics and Newton's mechanics that Maupertuis was led to this principle. He believed that Nature

† The principle was first mentioned in two papers, read (*a*) to the *Académie des Sciences de Paris* in 1744, and (*b*) to the Prussian Academy two years later.

acted by necessity in a manner which made *some* quantity a minimum, and a number of experimental results were plausibly explained if the product *m.v.s* were chosen as this quantity. Indeed, this product (fully defined) occurs in the exact formulation of the principle of least action; in consequence, it is not surprising that he found the observed data to be satisfactorily interpreted thereby. His derivation of *m.v.s* as the action was perhaps partly empirical. There is, further, strong evidence that he reached it from a somewhat vague conception of Leibniz' *vis viva* and the principle of virtual velocities.

The prime objective of Maupertuis will hardly be misrepresented if we state that he put forward the principle of least action in an endeavour to furnish not merely a rational but also a theological foundation for mechanics. His attempt at proving the existence of God through a physical law should be regarded as a last vestige of mediaeval scholasticism with its imperative to reconcile faith and reason. In his *Essai de cosmologie* (1759), he expounded the idea that the perfection of the Supreme Being in its Divine wisdom would be incompatible with anything other than utter simplicity and minimum expenditure of action.

> Notre principe, plus conforme aux idées que nous devons avoir des choses, laisse le Monde dans le besoin continuel de la puissance du Créateur, & est une suite nécessaire de l'emploi le plus sage de cette puissance. . . . Ces loix si belles & si simples sont peut-être les seules que le Créateur & l'Ordonnateur des choses a établies dans la matiere pour y opérer tous les phénomènes de ce Monde visible.[10]

It will be recalled that a similar outlook can be discerned in Leibniz' *Théodicée*.

The deficiencies in the principle as it was set forth by Maupertuis are illustrated in his examples.

(*a*) *The Direct Impact of Two Perfectly Inelastic Bodies.* Suppose the colliding bodies of masses m_1 and m_2 have initial velocities u_1 and u_2 respectively, the common final velocities being v. All velocities are measured in the same direction. In this problem, Maupertuis regards the distance as that described in unit time, i.e., as the velocity. The action thus becomes the product of the mass and the square of the velocity. He asserts that, in order to calculate the change of action, the relative velocities must be used and the above product summed for the two bodies. We thus have

$$\text{change of action} = m_1(u_1 - v)^2 + m_2(u_2 - v)^2.$$

Hence

$$m_1(u_1 - v)^2 + m_2(u_2 - v)^2 = \text{min.} \qquad . \qquad . \quad (7)$$

Differentiating, we get

$$m_1(u_1 - v)dv + m_2(u_2 - v)dv = 0. \qquad . \qquad . \qquad (8)$$

On cancellation of dv, we obtain the familiar law of the conservation of momentum, which determines v correctly.

(*b*) *The Direct Impact of Two Perfectly Elastic Bodies.* In this case, the two bodies will have different velocities after impact, say v_1 and v_2. Eq. (7) is therefore replaced by

$$m_1(u_1 - v_1)^2 + m_2(u_2 - v_2)^2 = \text{min.}, \qquad . \qquad (9)$$

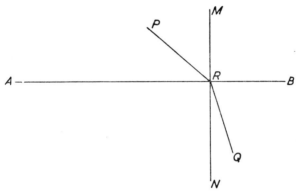

Fig. 6

and instead of (8) we may write

$$m_1(u_1 - v_1)dv_1 + m_2(u_2 - v_2)dv_2 = 0. \qquad . \qquad . \qquad (10)$$

Now, v_1 and v_2 are not independent, since the velocity of recession is equal to the velocity of approach, i.e.,

$$v_1 - v_2 = u_2 - u_1 = \text{const.} \qquad . \qquad . \qquad (11)$$

and hence $\qquad\qquad\qquad dv_1 = dv_2.$

Eq. (10) therefore becomes

$$m_1(u_1 - v_1) + m_2(u_2 - v_2) = 0$$

or

$$m_1u_1 + m_2u_2 = m_1v_1 + m_2v_2, \qquad . \qquad . \qquad (12)$$

which gives us once more the law of the conservation of momentum and which, together with (11), permits the computation of v_1 and v_2.

(*c*) *The Refraction of Light.* In Fig. 6, P denotes the initial point of the path of a light ray, Q the final point, AB a refracting surface, R the point of refraction and MN the normal to AB at R. v_1 and v_2 are the velocities of the light in the media above and below AB respectively. In this case, the action is calculated in still a different way, namely by multiplying the velocity of light in the medium by the total length of the path in that medium and adding the products in the two media,

i.e., $$v_1 . PR + v_2 . QR = \text{min.} \qquad . \qquad . \qquad (13)$$

The minimization of an expression of this form has been carried out in two previous examples; the result obtained was

$$\frac{\sin \angle PRM}{\sin \angle NRQ} = \frac{v_2}{v_1} = \text{const.} \qquad . \qquad . \qquad (14)$$

Snell's law is thus verified, but the true ratio of the velocities is, of course, the inverse of that obtained by Maupertuis.

From these instances it may immediately be inferred that Maupertuis had a diffuse idea, an intuition rather than a precise notion, of the action principle. One may, in fact, prefer to call it a conjecture and not a theorem. Due to this indefiniteness, Maupertuis' work should be ranked below that of Fermat, whose law exhibits mathematical coherence and consistency which do not admit of ambiguity. Even in its original form the postulate of least time can be appraised as an elegant though slightly inaccurate illustration of a minimum principle as we use the term today, whereas this estimate would scarcely be claimed for Maupertuis' discovery. Fermat was able, as already indicated, to derive correctly results which were previously unknown. In contrast, no contemporary physicist could fruitfully have employed the method of Maupertuis. His contribution was, however, of fundamental and far-reaching import in that it opened up a field of research that was soon to be explored and to lead to invaluable results.

Maupertuis' priority in the discovery of the principle of least action has not been uncontested. In 1757, the mathematician König produced a letter purported to have been written by Leibniz fifty years earlier, that contained a definite formulation of the principle. Reacting sharply to this claim, Maupertuis, president of the Prussian Academy, demanded that the original be demonstrated *ad oculos*—a request with which the Leibniz-scholar König was unable to comply. Maupertuis accused his fellow-member of plagiarism and denounced the document as a forgery. It may be not without a certain piquancy to note that Euler aligned himself with his French colleague in this controversy.

Although it is not within the scope of this monograph to discuss at length the evidence for or against the authenticity of the letter in question, it is worth recording that the first disclosure of this principle in the writings of Leibniz was made only well after it had already been proclaimed by another scientist. And furthermore, one cannot pay tribute to Leibniz for precedence in this matter on the basis of one spurious communication alone. Planck conceded the possibility of deception, but nevertheless accepted Helmholtz's arguments that Leibniz knew of the principle, and staunchly advocated this opinion by referring frequently to "Leibniz's principle of least action." While it is true that Leibniz anticipated, albeit in a vague, general manner, a law of that nature, his remarks barely warrant the assertion that he "knew" of it.

The whole dispute as to the priority of authorship has been unpleasantly tainted by personal feelings. The contingent fact that Maupertuis seems to have been of a strongly emotional personality, often arrogant and aggressive, must have aroused an ambivalent reaction towards him. Voltaire, for example, who once dubbed him "Sir Isaac Maupertuis," so elevating him to the lofty rank of Newton, later inveighed against his former idol with uncontrolled invective. National prejudices too may have played their part in the controversy. Regrettably, the annals of science reveal several occasions where patriotic bonds affected claims to antecedence and originality, as in the case of Boyle-Mariotte and Leibniz-Newton.

In conclusion, it is not surprising that most historians of natural philosophy uphold Maupertuis as the discoverer of the principle of least action, and we see no reason to depart from this standpoint.

§4

The Development of this Principle by Euler and Lagrange

THE principle of least action was first published as an exact dynamical theorem by Euler (1707–1793) in 1744, who confined himself to the particular case of a single particle moving in a plane curve.[11] His proposition asserted that, when a particle travels between two fixed points, it takes that path for which $\int vds$ is a minimum, v being the velocity of the particle and ds the corresponding element of the curve. "*Iam dico lineam a corpore descriptam ita fore comparatam, ut inter omnes alias lineas iisdem terminis contentas sit* $\int M \, ds\sqrt{v}$† *seu, ob M constans,* $\int ds\sqrt{v}$ *minimum.*" However, on analyzing Euler's argument, two reservations must be made to this formulation.

First, he based his demonstrations upon the calculus of variations—indeed, the principle was expounded in an addendum‡ to his classic memoir *Methodus inveniendi lineas curvas maximi minimive proprietate gaudentes, sive solutio problematis isoperimetrici latissimo sensu accepti*, where the fundamental methods of this calculus were evolved. Now this theory enables us to determine, not the curve between two fixed points for which a certain integral is a *minimum*, but that curve for which the difference between the integral taken along it, and the integral taken along any infinitely near curve between the two fixed points, is an infinitesimal quantity of the second order with respect to the distance between the curves. In other words, the integral along

† Euler denotes the square of the velocity by v.
‡ *De motu projectorum in medio non resistente, per methodum maximorum ac minimorum determinando.*

24

the determined curve is "stationary." We have already remarked in §2 that the above variational condition is necessary yet not sufficient for the integral to be a minimum; if the variational condition holds, the integral may be a minimum or a maximum or neither. Euler never considers the last possibility, though he sometimes speaks of the integral being a maximum or a minimum.

Further, he always reduces the problem to one of pure mathematics by computing, prior to an application of the calculus of variations, the particle's velocity as a function of its position. In more modern terms, he regards the energy of the particle as fixed. It is unequivocally mentioned in his treatise that the principle is only applicable to cases where the speed of the particle is dependent on its position alone, or, as we would say, to those cases where the forces are derivable from a potential and where the principle of the conservation of energy therefore holds. Nevertheless, he does not state in an explicit manner that the virtual path has to be subjected to the energy restriction.

Euler's principle may accordingly be reformulated as follows. A particle can travel between two fixed points with any given energy, and then moves in such a way that *the difference between the integral $\int vds$ taken along the real path and that taken along any neighbouring virtual† path between the two points is an infinitesimal quantity of second order; the particle is supposed to travel along the virtual path with the velocity for which the energy is equal to the given energy.* The condition is thus

$$\delta \int_P^Q vds = 0, \qquad . \qquad . \qquad . \qquad . \qquad (15)$$

where P and Q are the initial and final points, and δ denotes the variation of the integral under the aforementioned restrictions.

In his proof, Euler calculates the radius of curvature of the path directly, and also by means of the minimum principle; both operations, he observes, lead to the same result. Suppose the forces per unit mass on the particle in the x- and y-directions are X and Y respectively. The normal acceleration of the particle is then

$$\frac{X\dfrac{dy}{dx} - Y}{\sqrt{1 + \left(\dfrac{dy}{dx}\right)^2}}.$$

---

† By a virtual path is meant one along which the particle may be imagined to move without satisfying Newton's laws of motion.

Hence, if ρ is the radius of curvature, we have, from the ordinary formula,

$$\frac{v^2}{\rho} = \frac{Y - X\dfrac{dy}{dx}}{\sqrt{1 + \left(\dfrac{dy}{dx}\right)^2}}.$$

Turning to the variational method, we are required to minimize the integral

$$\int v\,ds = \int v\sqrt{1 + \left(\frac{dy}{dx}\right)^2}\,dx.$$

The Euler condition (see Appendix) for this integral to be stationary is

$$\frac{d}{dx}\left\{\frac{\partial}{\partial\left(\dfrac{dy}{dx}\right)}\left(v\sqrt{1 + \left(\frac{dy}{dx}\right)^2}\right)\right\} - \frac{\partial}{\partial y}\left(v\sqrt{1 + \left(\frac{dy}{dx}\right)^2}\right) = 0.$$

In carrying out the differentiation, we must employ the relations

$$\frac{\partial v^2}{\partial x} = 2X, \qquad \frac{\partial v^2}{\partial y} = 2Y,$$

since, in conformity with our previous discussion, the magnitude of the velocity has to be expressed as a function of the co-ordinates when the calculus of variations is applied. We therefore obtain

$$\frac{d}{dx}\left(v\,\frac{\dfrac{dy}{dx}}{\sqrt{1 + \left(\dfrac{dy}{dx}\right)^2}}\right) - \frac{Y}{v}\sqrt{1 + \left(\frac{dy}{dx}\right)^2} = 0,$$

or

$$\frac{\dfrac{dy}{dx}\left(X + Y\dfrac{dy}{dx}\right)}{v\sqrt{1 + \left(\dfrac{dy}{dx}\right)^2}} + \frac{v\,\dfrac{d^2y}{dx^2}}{\left\{1 + \left(\dfrac{dy}{dx}\right)^2\right\}^{3/2}} - \frac{Y}{v}\sqrt{1 + \left(\frac{dy}{dx}\right)^2} = 0,$$

as $\dfrac{dv^2}{dx} = 2\left(X + Y\dfrac{dy}{dx}\right)$. The last equation may be simplified to read

$$\frac{v^2 \dfrac{d^2y}{dx^2}}{\left\{1 + \left(\dfrac{dy}{dx}\right)^2\right\}^{3/2}} = \frac{Y - X\dfrac{dy}{dx}}{\sqrt{1 + \left(\dfrac{dy}{dx}\right)^2}},$$

whence, from the relation

$$\frac{1}{\rho} = \frac{\dfrac{d^2y}{dx^2}}{\left\{1 + \left(\dfrac{dy}{dx}\right)^2\right\}^{3/2}},$$

we get

$$\frac{v^2}{\rho} = \frac{Y - X\dfrac{dy}{dx}}{\sqrt{1 + \left(\dfrac{dy}{dx}\right)^2}}.$$

This equation agrees with the result found by the direct method.

Before dealing with this general problem, Euler had treated the particular instances of a particle, (*a*) in a uniform gravitational field, (*b*) under the action of combined constant horizontal and vertical forces, and (*c*) under the influence of a central force.

It need hardly be emphasized that Euler's principle differed basically from the conjecture of Maupertuis in that it presented us with a definite mathematical theorem which could be, and in fact was, developed into a most valuable branch of dynamical analysis. And yet his approach to the problem shows some signs of his predilection for *a priori* methods of reasoning and metaphysical speculation. His own introduction to the paper illustrates this tinge in his thought.

> Since all processes in nature obey certain maximum or minimum laws, there is no doubt that the curves, which bodies describe under the influence of arbitrary forces, also possess some maximum or minimum property. It does not seem so easy, however, to define this property *a priori* from metaphysical principles. But, as it is possible to determine the curves themselves with the aid of a direct method, one should be able, upon thorough examination of these curves, to conclude what quantity in them must be a maximum or minimum.

The first quantity to be considered for minimization, continues Euler, is that which results from the accelerating forces, i.e., it is the momentum, or more accurately, the aggregate of all the momenta. Thus he derives the integral $\int mv\,ds$. He infers that, once the calculations confirm this quantity to be a minimum, it will become easier to gain an insight into the concealed laws of nature and their final causes—*in intimas Naturae leges atque causas finales.*

At the end of his exposition, Euler maintains, though only on metaphysical grounds, that his minimum principle is also valid for the case of several particles.

Because in this instance the *motus* can only be expressed as a formula with difficulty, it is easier to comprehend and appreciate this theorem from fundamental principles than from the agreement of the calculations carried out according to the two methods. For, since the bodies resist every change of their state by reason of their inertia, they yield to the accelerating forces as little as possible, at least, if they are free. It therefore follows that in the actual motion the effect arising from the forces should be less than if the body or bodies were caused to move in any other manner. Although the force of this conclusion does not yet convince one as satisfactory, I do not doubt that it will be possible to justify it with the aid of a sane metaphysics. I leave this task, however, to others who are proficient in metaphysical studies.

These remarks, prompted by *a priori* considerations, are totally lacking in definiteness and rigidity, and it is perhaps surprising to find them in the work of one of the greatest mathematicians of all time.

Euler's manuscript was actually written in the second half of 1743, and we must hence assume that his discovery was made shortly before Maupertuis read his address on the principle of least action to the Paris Academy. On entering more deeply into the matter, however, we meet with the relevant fact that Maupertuis had broached the idea of maximum or minimum principles operative in nature as early as 1740.[12] He proved the theorem that bodies mutually attracting one another as the nth power of the distance between them are in equilibrium if, and only if, the sum of the products of the proportionality constants and the $(n + 1)^{th}$ power of the corresponding distances is a minimum. This is a well-known theorem which is a particular case of the law that a system is in equilibrium provided the potential energy is a minimum. In his most informative and lucid introduction to Euler's *Opera omnia*, Carathéodory records how Maupertuis had been in correspondence with Euler about the principle of least action. In a letter to Maupertuis dated 1745, he generously praised the paper of 1740 just mentioned, and, assuring him that he esteemed the maximum-minimum principle above his own mathematical researches devoted to the solution of particular problems, stated his conviction that some sort of maximum or minimum law prevails throughout nature. Later, there was a voluminous exchange of letters between the two scientists on this subject, and Euler identified his integral with Maupertuis' quantity of action and his theorem with Maupertuis' principle of least action. Withal, it appears that he was inspired, in part at least, by Maupertuis; and even Carathéodory, who is not reserved in his tribute to Euler's paper, is of the opinion that Maupertuis' vague contribution may have been the stimulus which led to the discovery made by the Swiss mathematician.

Although Euler was the first to implement Maupertuis' conjecture, the credit for having given the correct formulation of the principle of least action for general cases[13, 14] must be attributed to Lagrange (1736–1813). He treated what was in fact a system of mutually interacting particles, the forces between which were derived from a potential. The action of each particle is then defined as the integral $\int v\,ds$. Thence the principle adopts the form that *the system moves from one configuration to another in such a way as to make the total action, i.e., the sum of the actions of the individual particles, stationary as compared with adjacent virtual motions between the same two configurations and having the same energy as the actual motion.* Thus

$$\delta_{E \text{ const.}} \left(m_1 \int_A^B v_1 ds_1 + m_2 \int_A^B v_2 ds_2 + \ldots \right) = 0, \quad . \quad (16)$$

where m_i, v_i and ds_i denote the mass, the velocity and the element of the curve respectively of the ith particle, A the initial and B the final configuration.

It is perhaps worth noting that Lagrange simply stated that the expression $\sum m \int v\,ds$ is either a maximum or a minimum, without stressing the precise conditions which the virtual path must satisfy,† but his reasoning clearly implies the conditions

---

† To quote his original formulation: "*Soient tant de corps qu'on voudra M, M', M″, . . . , qui agissent les uns sur les autres d'une manière quelconque, et qui soient de plus, si l'on veut, animés par des forces centrales proportionnelles à des fonctions quelconques des distances; que s, s', s″, . . . , dénotent les espaces parcourus par ces corps dans le temps t, et que u, u', u″, . . . , soient leurs vitesses à la fin de ce temps; la formule*

$$M \int u\,ds + M' \int u'ds' + M'' \int u''ds'' + \ldots$$

sera toujours un maximum ou un minimum."

In the *Mécanique analytique*, however, he draws explicit attention to the two motions having the same initial and final configurations, but he still only implies the energy condition indirectly. He also extends the principle of least action to general dynamical systems.

just specified. A further point of interest, in passing, is that Maupertuis' original definition of action as the product of mass, velocity and distance occurs here again, but now in an exact form. If the method of Lagrange is applied to a system consisting of a single particle, the mass may be cancelled out and we obtain Euler's integral.

Lagrange's proof was substantially as follows—

Consider the motion of a system of particles between two configurations and construct any adjacent virtual path. Each point on the real path† is supposed to correspond to a point on the neighbouring path. In this and all subsequent proofs, the symbol d represents the differential of some quantity along one path, while the symbol δ stands for the variation of a quantity between any point on one path and the corresponding point on the adjacent path. Then, if x_i, y_i, z_i denote the co-ordinates of the ith particle w.r.t. a set of rectangular axes,

$$\delta\left(\sum_i m_i \int v_i ds_i\right) = \delta\left(\sum_i \sum_{x,y,z} m_i \int \dot{x}_i dx_i\right)$$

$$= \sum_i \sum_{x,y,z} \int \delta(m_i \dot{x}_i dx_i)$$

$$= \sum_i \sum_{x,y,z} \int \{m_i \dot{x}_i \delta(dx_i) + m_i \delta\dot{x}_i dx_i\}$$

$$= \sum_i \sum_{x,y,z} \int \{m_i \dot{x}_i d(\delta x_i) + m_i \delta\dot{x}_i dx_i\},$$

since $d(\delta x_i) = \delta(dx_i)$. Hence

$$\delta\left(\sum_i m_i \int v_i ds_i\right) = \sum_i \sum_{x,y,z} \int \{m_i d(\dot{x}_i \delta x_i) - m_i d\dot{x}_i \delta x_i + m_i \delta\dot{x}_i dx_i\}. \quad (17)$$

Now, because both paths have the same energy,

$\delta(\sum_i \sum_{x,y,z} \tfrac{1}{2}m_i \dot{x}_i{}^2 + V) = 0$, where V is the potential energy.

I.e., $$\sum_i \sum_{x,y,z} \left(m_i \dot{x}_i \delta\dot{x}_i + \frac{\partial V}{\partial x_i} \delta x_i\right) = 0. \quad . \quad . \quad (18)$$

---

† The word "path" is here to be interpreted as the continuous series of configurations through which the system moves.

Replacing dx_i in the third term of (17) by $\dot{x}_i dt$, and substituting

$$\sum_{x,y,z} \sum \left(-\frac{\partial V}{\partial x_i} \delta x_i \right) \text{ for } \sum_i \sum_{x,y,z} (m_i \dot{x}_i \delta \dot{x}_i),$$

we obtain

$$\delta \sum_i m_i \int v_i ds_i$$

$$= \sum_i \sum_{x,y,z} \int \left\{ m_i d(\dot{x}_i \delta x_i) - m_i d\dot{x}_i \delta x_i - \frac{\partial V}{\partial x_i} dt \delta x_i \right\}$$

$$= \sum_i \sum_{x,y,z} m_i \dot{x}_{if} \delta_f x_i - \int \sum_i \sum_{x,y,z} \left(m_i d\dot{x}_i + \frac{\partial V}{\partial x_i} dt \right) \delta x_i, \quad . \quad (19)$$

δ_f indicating the variation of a quantity between the configurations represented by the end-points of the paths, and $\dot{x}_{if}$ denoting the velocity at the end-point.

If the two paths are co-terminous, the first term in (19) disappears, and as the variations δx_i are arbitrary, the *necessary* and *sufficient* conditions for the variation of the action integral to vanish are that the coefficient of each δx_i must be equal to zero, i.e.,

$$m_i d\dot{x}_i + \frac{\partial V}{\partial x_i} dt = 0,$$

or

$$m_i \ddot{x}_i = -\frac{\partial V}{\partial x_i}. \quad . \quad . \quad (20)$$

A similar equation holds for each co-ordinate of each particle. Under these conditions, (19) gives directly

$$\delta \left(\sum_i m_i \int v_i ds_i \right) = 0. \quad . \quad . \quad (21)$$

This argument confirms that the principle of least action as expressed by Lagrange, together with the law of the conservation of energy, is fully equivalent to Newton's laws of motion and may, indeed, be employed as an alternative formulation of the principles of dynamics. Here, however, it will be regarded as a heuristic derivation from Newton's laws and as a mathematical device for obtaining further results. Lagrange himself characteristically remarked, in conformity with his general outlook on natural philosophy, that the principle of least action, as well as such theorems as the conservation of energy, were to be considered not as metaphysical postulates, but as simple and general consequences of the laws of mechanics. This attitude of mind differs fundamentally from that of Maupertuis and, to a lesser degree, from that of Euler and brings us nearer to a modern view of the laws of science.

For future discussion, it is desirable to compute the variation of the action integral between the real path and an adjacent virtual path not co-terminous with it, and along which the energy is not necessarily equal to the energy of the real path. Without postulating constancy of energy, eq. (18) becomes

$$\sum_i \sum_{x,y,z} \left(m_i \dot{x}_i \delta \dot{x}_i + \frac{\partial V}{\partial x_i} \delta x_i \right) = \delta E,$$

where E is the energy of the system.

Thus, (19) changes to

$$\delta \left(\sum_i m_i \int v_i ds_i \right)$$

$$= \sum_i \sum_{x,y,z} m_i \dot{x}_{if} \delta_f x_i + \int \left\{ - \sum_i \sum_{x,y,z} \left(m_i d\dot{x}_i + \frac{\partial V}{\partial x_i} dt \right) \delta x_i + \delta E dt \right\}$$

$$= \sum_i \sum_{x,y,z} m_i \dot{x}_{if} \delta_f x_i + \int \delta E dt, \qquad . \qquad . \qquad . \qquad . \qquad (22)$$

the first terms under the integral sign disappearing by reason of Newton's laws. If neither the end-points nor the initial points of the paths coincide, (22) must be modified to

$$\delta \left(\sum_i m_i \int v_i ds_i \right) = \sum_i \sum_{x,y,z} (m_i \dot{x}_{if} \delta_f x_i - m_i \dot{x}_{ia} \delta_a x_i) + \int \delta E dt, \quad . \quad (23)$$

where the suffix a refers to the beginning of the path. For the case in which the paths are co-terminous and have the same energy, the right-hand side of (23) vanishes, and the equation reduces to (21).

For the next half century, the principle of least action was thought of as interesting rather than important, and no use at all was made of it. Planck sarcastically recalls how Poisson, in 1837, described it as "only a useless rule." So it was not before Hamilton that any further work was devoted to action principles.

§5

The Equations of Lagrange and Hamilton

NEWTON'S mechanics was founded upon the concept of point masses, and his equations of motion were stated in terms of the rectangular Cartesian co-ordinates of the particles. While the problems of dynamics can theoretically be solved by such means— rigid bodies being treated as an assemblage of an infinite number of particles—the integration of the equations involved is generally very complicated. Special methods had therefore to be developed in order to arrive at the desired results. The starting point of these more adequate methods is to be found in the equations of motion of Lagrange.[14] He considered a system to be specified by means of "generalized co-ordinates," i.e., any set of variables sufficient in number to define unambiguously the configuration of the system. By utilizing the expressions for the kinetic and potential energies as functions of these co-ordinates, he succeeded in expressing the equations of motion in such a way as to be applicable in the same form whichever generalized co-ordinates are chosen. In his own words: ". . . it is perhaps one of the principal advantages of our method that it expresses the equations of every problem in the most simple form relative to each set of variables and that it enables us to see beforehand which variables one should use in order to facilitate the integration as much as possible." These equations of Lagrange will be used to extend the range of validity of the action principle to generalized co-ordinates. Moreover, the developments to which they gave rise are intimately connected with variational principles. A brief digression is therefore necessary to develop these equations and the canonical equations of Hamilton which are derived from them.

Suppose the system requires n generalized co-ordinates to

specify its configuration and denote these generalized co-ordinates by $q_1, q_2, \ldots, q_n$. The rectangular co-ordinates x_i, y_i, z_i of one of the particles or of an element of one of the rigid bodies will then be functions of the q's. It usually occurs that the total number of q's is less than the total number of x's, y's, and z's, because the system will probably be subjected to certain kinematical restrictions, e.g., that some of the bodies within the system are rigid. These restrictions will be regarded as being brought about by "internal forces." Throughout this discussion, our attention will be confined to "holonomic" systems, i.e., to those in which the restrictions imposed on small displacements are integrable. This stipulation does not, however, hold universally. In a system composed of a sphere which rolls without slipping on a plane, for instance, only three independent infinitesimal displacements are possible, but all five generalized co-ordinates are nevertheless necessary to specify the system.

We now introduce the concept of "generalized force." The work done by the system in an infinitesimal displacement will be proportional to the elements dq_r of the generalized co-ordinates through which it is displaced. Let these proportionality factors be represented by Q_r, i.e.,

$$\text{work done} = \sum_r Q_r dq_r. \qquad . \qquad . \qquad (24)$$

Q_r is then defined as the generalized force corresponding to the co-ordinate q_r. We need only take into account the work done by the external forces, since the work done by the internal forces is zero. From this definition it follows that the generalized force can be expressed as a function of the Newtonian forces acting on the individual particles and elements of the rigid bodies belonging to the system. For, setting X_i equal to the x-force on the ith particle, etc., we have, summing over the whole system,

$$\text{work done} = \sum_i \sum_{x,y,z} X_i dx_i$$

$$= \sum_i \sum_{x,y,z} X_i \sum_r \frac{\partial x_i}{\partial q_r} dq_r$$

$$= \sum_r \left(\sum_i \sum_{x,y,z} X_i \frac{\partial x_i}{\partial q_r} \right) dq_r. \qquad . \qquad (25)$$

Comparing (24) and (25),

$$Q_r = \sum_i \sum_{x,y,z} X_i \frac{\partial x_i}{\partial q_r}. \qquad . \qquad . \qquad (26)$$

This formula for Q_r enables us to transform from Newton's laws to Lagrange's equations.† Thus, putting $X_i = m_i\ddot{x}_i$ in (26), where m_i is the mass of the ith particle, we get

$$Q_r = \sum_i \sum_{x,y,z} m_i\ddot{x}_i \frac{\partial x_i}{\partial q_r}$$

$$= \sum_i \sum_{x,y,z} m_i \frac{d\dot{x}_i}{dt} \frac{\partial x_i}{\partial q_r}$$

$$= \sum_i \sum_{x,y,z} \left\{ m_i \frac{d}{dt}\left(\dot{x}_i \frac{\partial x_i}{\partial q_r}\right) - m_i\dot{x}_i \frac{d}{dt}\left(\frac{\partial x_i}{\partial q_r}\right) \right\}$$

$$= \sum_i \sum_{x,y,z} \left\{ m_i \frac{d}{dt}\left(\dot{x}_i \frac{\partial x_i}{\partial q_r}\right) - m_i\dot{x}_i \sum_s \frac{\partial}{\partial q_s}\left(\frac{\partial x_i}{\partial q_r}\right)\dot{q}_s \right\}$$

$$= \sum_i \sum_{x,y,z} \left\{ m_i \frac{d}{dt}\left(\dot{x}_i \frac{\partial x_i}{\partial q_r}\right) - m_i\dot{x}_i \sum_s \frac{\partial}{\partial q_r}\left(\frac{\partial x_i}{\partial q_s}\right)\dot{q}_s \right\}$$

$$= \sum_i \sum_{x,y,z} \left\{ m_i \frac{d}{dt}\left(\dot{x}_i \frac{\partial x_i}{\partial q_r}\right) - m_i\dot{x}_i \sum_s \frac{\partial}{\partial q_r}\left(\frac{\partial x_i}{\partial q_s}\dot{q}_s\right) \right\}$$

$$\left(\text{since } \frac{\partial \dot{q}_s}{\partial q_r} = 0\right)$$

$$= \sum_i \sum_{x,y,z} \left\{ m_i \frac{d}{dt}\left(\dot{x}_i \frac{\partial x_i}{\partial q_r}\right) - m_i\dot{x}_i \frac{\partial \dot{x}_i}{\partial q_r} \right\}.$$

Now, $\quad \dfrac{\partial \dot{x}_i}{\partial \dot{q}_r} = \dfrac{\partial}{\partial \dot{q}_r}\sum_s \dfrac{\partial x_i}{\partial q_s}\dot{q}_s$

$$= \sum_s \frac{\partial x_i}{\partial q_s}\frac{\partial \dot{q}_s}{\partial \dot{q}_r} \quad \left(\frac{\partial x_i}{\partial q_s} \text{ being independent of the } \dot{q}\text{'s}\right)$$

$$= \frac{\partial x_i}{\partial q_r}, \qquad . \qquad . \qquad . \qquad . \qquad . \qquad . \qquad (27)$$

since $\quad \dfrac{\partial \dot{q}_s}{\partial \dot{q}_r} = 0 \quad (s \neq r)$

$$= 1 \quad (s = r).$$

Therefore $\quad Q_r = \sum_i \sum_{x,y,z} \left\{ m_i \frac{d}{dt}\left(\dot{x}_i \frac{\partial \dot{x}_i}{\partial \dot{q}_r}\right) - m_i\dot{x}_i \frac{\partial \dot{x}_i}{\partial q_r} \right\}$

$$= \frac{d}{dt}\left\{ \frac{\partial}{\partial \dot{q}_r}\left(\sum_i \sum_{x,y,z} \tfrac{1}{2}m_i\dot{x}_i^2\right) \right\} - \frac{\partial}{\partial q_r}\left(\sum_i \sum_{x,y,z} \tfrac{1}{2}m_i\dot{x}_i^2\right).$$

† The proof is essentially the same as that given by Hamilton.[15]

But $\sum_i \sum_{x,y,z} \frac{1}{2} m_i \dot{x}_i^2$ is a familiar quantity, namely the kinetic energy of the system, for which the symbol T will be used. Hence we obtain finally

$$\frac{d}{dt}\left(\frac{\partial T}{\partial \dot{q}_r}\right) - \frac{\partial T}{\partial q_r} = Q_r. \qquad . \qquad . \qquad . \qquad (28)$$

Equations (28)—one for each generalized co-ordinate—are the Lagrange equations of the system. Before they can be applied, T and the Q's must be calculated in terms of the generalized co-ordinates.

In all processes in which the principle of the conservation of energy holds, the forces can be derived from a potential, that is, there exists a function V of the co-ordinates such that the work done *on* the system in changing from the configuration (q_{1a}, q_{2a}, . . . , q_{na}) to the configuration (q_{1b}, q_{2b}, . . . , q_{nb}) is equal to V (q_{1b}, q_{2b}, . . . , q_{nb}) $- V$ ($q_{1a}, q_{2a}, . . . , q_{na}$). The work done *by* the system in an infinitesimal displacement is therefore given by the equation

$$\text{work done} = -\sum_r \frac{\partial V}{\partial q_r} dq_r. \qquad . \qquad . \qquad (29)$$

A comparison between (24) and (29) shows immediately that

$$Q_r = -\frac{\partial V}{\partial q_r}. \qquad . \qquad . \qquad . \qquad (30)$$

A second form of the Lagrange equations (28), involving the potential energy V in place of the generalized forces, is therefore

$$\frac{d}{dt}\left(\frac{\partial T}{\partial \dot{q}_r}\right) - \frac{\partial T}{\partial q_r} = -\frac{\partial V}{\partial q_r}. \qquad . \qquad . \qquad (31)$$

The equations of Lagrange can further be written in still another way employing a single function, the "Lagrangian," defined by

$$L = T - V. \qquad . \qquad . \qquad . \qquad (32)$$

As $\partial V/\partial \dot{q}_r = 0$, equation (31) then becomes

$$\boxed{\frac{d}{dt}\left(\frac{\partial L}{\partial \dot{q}_r}\right) - \frac{\partial L}{\partial q_r} = 0} \qquad . \qquad . \qquad . \qquad (33)$$

This formulation of Lagrange's equations is the most powerful for theoretical purposes and will be frequently invoked in subsequent discussions.

The kinetic energy T, being a quadratic function of the $\dot{x}$'s, is also a quadratic function of the $\dot{q}$'s, which are linearly connected with the $\dot{x}$'s by means of the relations

$$dx_i = \sum_r \frac{\partial x_i}{\partial q_r} dq_r,$$

i.e.,
$$\frac{dx_i}{dt} = \sum_r \frac{\partial x_i}{\partial q_r} \frac{dq_r}{dt}. \qquad \qquad (34)$$

From a well-known and easily demonstrated theorem of Euler, it follows that

$$2T = \sum_r \frac{\partial T}{\partial \dot{q}_r} \dot{q}_r \qquad \cdots \qquad (35)$$

or, since $\qquad \partial V/\partial \dot{q}_r = 0,$

$$2T = \sum_r \frac{\partial L}{\partial \dot{q}_r} \dot{q}_r. \qquad \cdots \qquad (36)$$

This equation leads, without difficulty, to the principle of the conservation of energy: by simple derivation we obtain

$$\frac{d}{dt}(T + V) = \frac{d}{dt}(2T - L)$$

$$= \frac{d(2T)}{dt} - \sum_r \frac{\partial L}{\partial q_r} \dot{q}_r - \sum_r \frac{\partial L}{\partial \dot{q}_r} \ddot{q}_r$$

$$= \frac{d(2T)}{dt} - \sum_r \frac{d}{dt}\left(\frac{\partial L}{\partial \dot{q}_r}\right) \dot{q}_r - \sum_r \frac{\partial L}{\partial \dot{q}_r} \ddot{q}_r,$$

$$\text{(by Lagrange's equations)}$$

$$= \frac{d}{dt}\left(\sum_r \frac{\partial L}{\partial \dot{q}_r} \dot{q}_r\right) - \frac{d}{dt}\left(\sum_r \frac{\partial L}{\partial \dot{q}_r} \dot{q}_r\right)$$

$$= 0. \qquad \cdots \qquad (37)$$

From this it follows that $T + V$ remains constant with the motion.

The most significant property of equations of the form (33) is their invariance with respect to arbitrary co-ordinate transformations. The validity of the Lagrange equations in generalized co-ordinates follows, indeed, at once, from this and the immediately evident fact that they hold in rectangular co-ordinates; rigid bodies must here be treated as systems of mutually attracting particles. For future reference, it is necessary to prove that, if the equations are true in the co-ordinate system q_r, they are equally valid for any other co-ordinate system Q_s,† where the Q's (n in number) are any functions of the q's and t. The Q's must, of course, be defined so as to determine the system completely. L can be any function of the q's, $\dot{q}$'s and t. From the usual formula for transformation of derivatives,

$$\frac{d}{dt}\left(\frac{\partial L}{\partial \dot{Q}_s}\right) - \frac{\partial L}{\partial Q_s} = \sum_r \left\{ \frac{d}{dt}\left(\frac{\partial L}{\partial \dot{q}_r}\frac{\partial \dot{q}_r}{\partial \dot{Q}_s} + \frac{\partial L}{\partial q_r}\frac{\partial q_r}{\partial \dot{Q}_s} + \frac{\partial L}{\partial t}\frac{\partial t}{\partial \dot{Q}_s}\right) \right.$$

$$\left. - \frac{\partial L}{\partial q_r}\frac{\partial q_r}{\partial Q_s} - \frac{\partial L}{\partial \dot{q}_r}\frac{\partial \dot{q}_r}{\partial Q_s} - \frac{\partial L}{\partial t}\frac{\partial t}{\partial Q_s} \right\}$$

$$= \sum_r \left\{ \frac{d}{dt}\left(\frac{\partial L}{\partial \dot{q}_r}\frac{\partial q_r}{\partial Q_s}\right) - \frac{\partial L}{\partial q_r}\frac{\partial q_r}{\partial Q_s} - \frac{\partial L}{\partial \dot{q}_r}\frac{\partial \dot{q}_r}{\partial Q_s} \right\}$$

$$\left(\text{as } \frac{\partial q_r}{\partial \dot{Q}_s} = 0, \quad \frac{\partial t}{\partial \dot{Q}_s} = 0, \quad \frac{\partial t}{\partial Q_s} = 0\right.$$

$$\left.\text{and, analogously to (27), } \frac{\partial \dot{q}_r}{\partial \dot{Q}_s} = \frac{\partial q_r}{\partial Q_s}\right)$$

$$= \sum_r \left\{ \frac{d}{dt}\left(\frac{\partial L}{\partial \dot{q}_r}\right)\frac{\partial q_r}{\partial Q_s} + \frac{\partial L}{\partial \dot{q}_r}\frac{d}{dt}\left(\frac{\partial q_r}{\partial Q_s}\right) - \frac{\partial L}{\partial q_r}\frac{\partial q_r}{\partial Q_s} \right.$$

$$\left. - \frac{\partial L}{\partial \dot{q}_r}\frac{\partial}{\partial Q_s}\left(\frac{dq_r}{dt}\right) \right\}$$

$$= \sum_r \left\{ \frac{d}{dt}\left(\frac{\partial L}{\partial \dot{q}_r}\right)\frac{\partial q_r}{\partial Q_s} + \frac{\partial L}{\partial \dot{q}_r}\sum_u\left(\frac{\partial}{\partial Q_u}\frac{\partial q_r}{\partial Q_s}\dot{Q}_u + \frac{\partial}{\partial t}\frac{\partial q_r}{\partial Q_s}\right) \right.$$

$$\left. - \frac{\partial L}{\partial q_r}\frac{\partial q_r}{\partial Q_s} - \frac{\partial L}{\partial \dot{q}_r}\sum_u\frac{\partial}{\partial Q_s}\left(\frac{\partial q_r}{\partial Q_u}\dot{Q}_u + \frac{\partial q_r}{\partial t}\right) \right\}$$

$$= \sum_r \left\{ \frac{d}{dt}\left(\frac{\partial L}{\partial \dot{q}_r}\right)\frac{\partial q_r}{\partial Q_s} + \frac{\partial L}{\partial \dot{q}_r}\sum_u\left(\dot{Q}_u\frac{\partial^2 q_r}{\partial Q_u\partial Q_s} + \frac{\partial^2 q_r}{\partial Q_s\partial t}\right) - \frac{\partial L}{\partial q_r}\frac{\partial q_r}{\partial Q_s} \right.$$

$$\left. - \frac{\partial L}{\partial \dot{q}_r}\sum_u\left(\dot{Q}_u\frac{\partial^2 q_r}{\partial Q_u\partial Q_s} + \frac{\partial^2 q_r}{\partial Q_s\partial t}\right) \right\}$$

$$= \sum_r \left\{ \frac{d}{dt}\left(\frac{\partial L}{\partial \dot{q}_r}\right) - \frac{\partial L}{\partial q_r} \right\}\frac{\partial q_r}{\partial Q_s}$$

$$= 0. \qquad . \qquad . \qquad . \qquad . \qquad . \qquad . \qquad . \qquad . \qquad (38)$$

† These Q's have, of course, to be distinguished from those denoting the generalized forces, although the same symbols are used.

A slight modification of the argument at the end of this proof shows that equations of the type (31), where V is a function of the q's and t and not of the $\dot{q}$'s, remain invariant with respect to these co-ordinate transformations.

Likewise, an expression of the form $\sum_r \dfrac{\partial L}{\partial \dot{q}_r} \dot{q}_r$ is invariant with respect to similar co-ordinate transformations which do not involve the time. For, transforming the derivatives in the customary way,

$$
\begin{aligned}
\sum_s \frac{\partial L}{\partial \dot{Q}_s} \dot{Q}_s &= \sum_s \left\{ \left(\sum_r \frac{\partial L}{\partial \dot{q}_r} \frac{\partial q_r}{\partial Q_s} \right) \left(\sum_u \frac{\partial Q_s}{\partial q_u} \dot{q}_u \right) \right\} \\
&= \sum_r \sum_u \frac{\partial L}{\partial \dot{q}_r} \dot{q}_u \left(\sum_s \frac{\partial q_r}{\partial Q_s} \frac{\partial Q_s}{\partial q_u} \right) \\
&= \sum_r \sum_u \frac{\partial L}{\partial \dot{q}_r} \dot{q}_u \frac{\partial q_r}{\partial q_u} \\
&= \sum_r \frac{\partial L}{\partial \dot{q}_r} \dot{q}_r. \qquad \cdot \quad \cdot \quad \cdot \quad \cdot \quad (39)
\end{aligned}
$$

We hereby obtain an alternative proof of (35) and (36), because these expressions for the kinetic energy are valid, as is evident on inspection, in rectangular co-ordinate systems.

Hitherto we have restricted ourselves to classical, i.e., non-relativistic mechanics. Lagrange's equations remain true in the theory of relativity, though only for the case of a single particle moving in a field of force derivable from a potential. The function L must, however, be determined differently, since it is patent that eq. (31), interpreted relativistically, does not hold in rectangular co-ordinates as it stands. This equation will be satisfied provided we are able to find some function F, for which

$$
\frac{\partial F}{\partial \dot{x}} = x\text{-momentum} = \frac{m_0 \dot{x}}{\sqrt{1 - \dfrac{v^2}{c^2}}}. \qquad \cdot \qquad (40)
$$

Such a function is furnished by

$$
F = m_0 c^2 \left(1 - \sqrt{1 - \frac{v^2}{c^2}} \right). \qquad \cdot \qquad (41)
$$

Hence, from (40) and the trivial equation $\partial F / \partial x = 0$, we get

$$\frac{d}{dt}\left(\frac{\partial F}{\partial \dot{x}}\right) - \frac{\partial F}{\partial x} = -\frac{\partial V}{\partial x}. \qquad . \qquad . \qquad (42)$$

Similar equations hold for the co-ordinates y and z. By virtue of the invariance property of (42), it follows at once that, if q_r ($r = 1, 2, 3$) are a set of generalized co-ordinates,

$$\frac{d}{dt}\left(\frac{\partial F}{\partial \dot{q}_r}\right) - \frac{\partial F}{\partial q_r} = -\frac{\partial V}{\partial q_r}, \qquad (r = 1, 2, 3) \qquad (43)$$

which are the Lagrange equations for relativity mechanics. Let the relativistic Lagrange function be defined by

$$L = F - V; \qquad . \qquad . \qquad . \qquad . \qquad (44)$$

eq. (33) is then equivalent to (43), as in the classical case.

Equation (36) must in relativistic mechanics be substituted by

$$T + F = \sum_r \frac{\partial F}{\partial \dot{q}_r}\,\dot{q}_r = \sum \frac{\partial L}{\partial \dot{q}_r}\,\dot{q}_r, \qquad . \qquad . \qquad (45)$$

where T represents the kinetic energy $m_0 c^2 \left(\dfrac{1}{\sqrt{1 - \dfrac{v^2}{c^2}}} - 1\right)$.

The verification of (45) for rectangular co-ordinates is straight-forward, and the equation can, from its invariance property, be generalized directly to arbitrary co-ordinate systems. On writing $T + F$ for $2T$, wherever it occurs in the proof of (37), the law of the conservation of energy can be established as before.

The solution of a dynamical problem by Lagrange's method requires the integration of n second-order differential equations in the n unknowns $q_1, q_2, \ldots, q_n$. An alternative system, proposed by Hamilton (1805–1865) and bearing his name, consists of $2n$ first-order differential equations in $2n$ unknowns, and has the advantage that it is extremely simple and concise in its formulation.[15] Hamilton's equations, the so-called "canonical" equations, constitute the basis of the advanced theory of dynamics.

An original concept introduced by Hamilton is the "generalized momentum," which is defined by

$$p_r = \frac{\partial L}{\partial \dot{q}_r}, \qquad \cdots \qquad (46)$$

an expression that has already been encountered on previous occasions. If the q's are rectangular co-ordinates, the generalized momenta reduce to the ordinary momenta. That the generalized momenta are uniquely determined by the velocities is a trivial observation; the converse, it must be stressed, also holds. To show this, we need merely note that, in classical mechanics, T is a homogeneous quadratic function of the $\dot{q}$'s, and knowledge of the n momenta affords us n linear equations for the n $\dot{q}$'s. As T is a positive definite quadratic form in the $\dot{q}$'s, the determinant derived from the coefficients in these equations cannot vanish. Our assertion is thus proved. In the relativistic mechanics of a single particle, the velocities are uniquely determined by the momenta in a rectangular co-ordinate system, and the same result therefore follows by co-ordinate transformation in a generalized co-ordinate system. The generalized co-ordinates and momenta may consequently be employed as alternative to the generalized co-ordinates and velocities in order to specify a system with regard to the positions and velocities of its elements.

We next define the "Hamiltonian function" H as the energy of the system expressed as a function of the generalized *co-ordinates* and *momenta*. This function is of fundamental significance and will frequently occur in the following chapters. In classical mechanics we find

$$H = T + V$$
$$= 2T - (T - V)$$
$$= \sum_r p_r \dot{q}_r - L, \qquad \text{by (35),} \qquad . \quad (47)$$

and in relativistic mechanics,

$$H = T + V$$
$$= (T + F) - (F - V)$$
$$= \sum_r p_r \dot{q}_r - L, \text{ as in the classical case.}$$

After these preliminary results, we can now proceed to derive Hamilton's canonical equations.

From (47),

$$\frac{\partial H}{\partial p_r} \equiv \left(\frac{\partial H}{\partial p_r}\right)_q \quad \text{(the suffix } q \text{ indicating that the } q\text{'s are to be kept constant)}$$

$$= \dot{q}_r + \sum_s p_s \left(\frac{\partial \dot{q}_s}{\partial p_r}\right)_q - \left(\frac{\partial L}{\partial p_r}\right)_q$$

$$= \dot{q}_r + \sum_s \left(\frac{\partial L}{\partial \dot{q}_s}\right)_q \left(\frac{\partial \dot{q}_s}{\partial p_r}\right)_q - \sum_s \left(\frac{\partial L}{\partial \dot{q}_s}\right)_q \left(\frac{\partial \dot{q}_s}{\partial p_r}\right)$$

$$= \dot{q}_r,$$

and similarly

$$\frac{\partial H}{\partial q_r} \equiv \left(\frac{\partial H}{\partial q_r}\right)_p$$

$$= \sum_s p_s \left(\frac{\partial \dot{q}_s}{\partial q_r}\right)_p - \left(\frac{\partial L}{\partial q_r}\right)_p$$

$$= \sum_s p_s \left(\frac{\partial \dot{q}_s}{\partial q_r}\right)_p - \left(\frac{\partial L}{\partial q_r}\right)_{\dot{q}} - \sum_s \left(\frac{\partial L}{\partial \dot{q}_s}\right)_q \left(\frac{\partial \dot{q}_s}{\partial q_r}\right)_p$$

$$= \sum_s p_s \left(\frac{\partial \dot{q}_s}{\partial q_r}\right)_p - \frac{d}{dt}\left(\frac{\partial L}{\partial \dot{q}_r}\right)_q - \sum_s \left(\frac{\partial L}{\partial \dot{q}_s}\right)_q \left(\frac{\partial \dot{q}_s}{\partial q_r}\right)_p,$$

from Lagrange's equations

$$= \sum_s p_s \left(\frac{\partial \dot{q}_s}{\partial q_r}\right)_p - \frac{dp_r}{dt} - \sum_s p_s \left(\frac{\partial \dot{q}_s}{\partial q_r}\right)_p,$$

from the definition of the p's,

$$= -\dot{p}_r.$$

Rewriting the results

$$\left.\begin{array}{ll} (a) & \dfrac{\partial H}{\partial p_r} = \dot{q}_r \\[2em] (b) & \dfrac{\partial H}{\partial q_r} = -\dot{p}_r \end{array}\right\}, \quad \cdot \quad \cdot \quad \cdot \quad (48)$$

we have the canonical equations of motion for a dynamical system. The first set of (48) depends on the definition of H

alone, whereas the second is a consequence of Lagrange's equations of motion.

The converse of (48a) is valid in the sense that, if we are given quantities p_r which satisfy these equations, H being the same function of the p's and q's as before, these quantities p_r will be equal to $\partial L/\partial \dot{q}_r$. To this end, denote provisionally the given quantities that satisfy (48a) by p_r', and set the ordinary momenta equal to p_r. As the $\dot{q}_r$'s are determined by the p's, we may write

$$\dot{q}_r = f_r(q_s,p_s).$$

The derivatives of H are consequently

$$\frac{\partial H(q_s,p_s)}{\partial p_r} \equiv f_r(q_s,p_s),$$

or

$$\frac{\partial H(q_s,p_s')}{\partial p_r'} \equiv f_r(q_s,p_s').$$

But we are given

$$\frac{\partial H(q_s,p_s')}{\partial p_r'} = \dot{q}_r.$$

Therefore

$$\dot{q}_r = f_r(q_s,p_s').$$

Hence the $\dot{q}_r$'s can be evaluated from the p_r''s in exactly the same way as they are obtained from the p_r's. If, therefore, we compute the $\dot{q}_r$'s from the p_r''s and thereupon calculate the p_r's from the $\dot{q}_r$'s, we shall get

$$p_r = p_r',$$

i.e.,

$$p_r' = \frac{\partial L}{\partial \dot{q}_r}.$$

It now follows that the system is moving in accordance with the ordinary laws of dynamics if we can find a set of p's for which (48) holds.

For the actual solution of problems, the equations of Lagrange are more convenient than those of Hamilton, since the first step in integrating Hamilton's equations would amount to reducing their number by half, an operation which would lead us back to our original Lagrange equations. In purely theoretical inquiries, on the other hand, Hamilton's equations are often more useful.

The dominant position of the equations of Lagrange in the history of dynamics cannot be better illustrated than by citing Hamilton's own words. "The theoretical development of the laws of motion of bodies is of such interest and importance, that it has engaged the attention of all the most eminent mathematicians, since the invention of dynamics as a mathematical science by Galileo. . . . Among the successors of those illustrious men, Lagrange has perhaps done more than any other analyst, to give extent and harmony to such deductive researches, by showing that the most varied consequences respecting the motions of systems of bodies may be derived from one radical formula; the beauty of the method so suiting the dignity of the results, as to make his great work a kind of scientific poem."[16]

§6

Hamilton's Principle and the Hamilton-Jacobi Equation

HAVING developed these methods of Lagrange and Hamilton, we are now equipped to give a more general treatment of variational principles.

Our first objective is the extension of the principle of least action, as expressed in (21), to generalized co-ordinate systems. The principle then adopts the form

$$\delta \int \sum_r p_r dq_r = 0. \qquad \qquad \qquad (49)$$

Eq. (49) is in fact equivalent to (21) by reason of the invariance of the integrand, which can be verified as in (39). The general form of the principle may also be derived directly with the aid of Lagrange's equations in the following manner—

$$\delta \int \sum_r p_r dq_r = \int \sum_r \delta \left(\frac{\partial L}{\partial \dot{q}_r} dq_r \right)$$

$$= \int \sum_r \left\{ \delta \left(\frac{\partial L}{\partial \dot{q}_r} \right) dq_r + \frac{\partial L}{\partial \dot{q}_r} \delta(dq_r) \right\}$$

$$= \int \sum_r \left\{ \delta \left(\frac{\partial L}{\partial \dot{q}_r} \right) \dot{q}_r dt + d \left(\frac{\partial L}{\partial \dot{q}_r} \delta q_r \right) - d \left(\frac{\partial L}{\partial \dot{q}_r} \right) \delta q_r \right\}$$

$$= \int \sum_r \left\{ \delta \left(\frac{\partial L}{\partial \dot{q}_r} \right) \dot{q}_r dt + d \left(\frac{\partial L}{\partial \dot{q}_r} \delta q_r \right) - \frac{\partial L}{\partial q_r} \delta q_r dt \right\},$$

by Lagrange's equations.

From (47) we obtain

$$\delta H = \sum_r \left\{ \dot{q}_r \delta \left(\frac{\partial L}{\partial \dot{q}_r} \right) + \frac{\partial L}{\partial \dot{q}_r} \delta \dot{q}_r - \frac{\partial L}{\partial q_r} \delta q_r - \frac{\partial L}{\partial \dot{q}_r} \delta \dot{q}_r \right\}$$

$$= \dot{q}_r \delta \left(\frac{\partial L}{\partial \dot{q}_r} \right) - \frac{\partial L}{\partial q_r} \delta q_r.$$

Hence $\delta \int \sum_r p_r dq_r = \int \left\{ \sum_r d \left(\frac{\partial L}{\partial \dot{q}_r} \delta q_r \right) + \delta H dt \right\}$

$$= \sum_r (p_{rf} \delta_f q_r - p_{ra} \delta_a q_r) + \int \delta H dt, \quad . \quad . \quad . \quad (50)$$

where the a's and f's denote as before the momenta and co-ordinate variations at the beginning and end of the path respectively. In the case where the real and virtual paths are co-terminous and have the same energy, (50) reduces to (49).

The expression $\sum_r p_r dq_r$ in (50) may be written $\sum_r p_r \dot{q}_r dt$, which is equal to $2Tdt$ (or $(T + F)dt$ in relativity theory). Hence an alternative formulation of eq. (50) is

$$\delta \int 2T dt = \sum_r (p_{rf} \delta_f q_r - p_{ra} \delta_a q_r) + \int \delta H dt \text{ (in classical theory)}$$

$$\delta \int (T + F) dt = \sum_r \delta(p_f q_r - p_{ra} \delta_a q_r) + \int \delta H dt \text{ (in relativity theory)}$$

$$\biggr\},(51)$$

and, when the two paths are co-terminous and have the same energy, the corresponding formulation of (49) becomes

$$\delta \int 2T dt = 0 \text{ (classical)}$$
$$\delta \int (T + F) dt = 0 \text{ (relativistic)}$$

$$\biggr\} \quad . \quad . \quad . \quad (52)$$

The first of these equations was called by Lagrange the "principle of smallest or greatest living force."

In its original form, as stated by Lagrange, the principle of least action suffers from the limitation that it applies only to virtual paths having the same energy as the real path. Accordingly, the principle depends, in its formulation, on the law of the conservation of energy, which must therefore be regarded as logically prior to it. In all practical computations or theoretical problems making use of the action principle, the energy condition must first be applied. This sometimes results in certain difficulties and awkwardness in derivation. By removing this restriction concerning the energy of the virtual path, Hamilton enunciated the principle in what is generally recognized to be its most effectual form.[16]

Hamilton considered the variation of $\int L dt$, which may be deduced from (51) in the following manner—

$$\delta \int L dt = \begin{cases} \int (2T - H) dt \text{ (in classical theory)} \\ \int (T + F - H) dt \text{ (in relativistic theory)} \end{cases}$$

$$= \sum_r (p_{rf} \delta_f q_r - p_{ra} \delta_a q_r) + \int \delta H dt - \delta \int H dt.$$

The variation of the last integral arises from two different sources—

(i) a quantity $\int \delta H dt$ due to the variation of energy between the real and virtual paths, and

(ii) a quantity $(H_f \delta_f t - H_a \delta_a t)$ originating from the time variation between the two paths at their terminal points. Hence we can write

$$\delta \int L dt = \sum_r (p_{rf} \delta_f q_r - p_{ra} \delta_a q_r) + \int \delta H dt - \int \delta H dt - (H_f \delta_f t - H_a \delta_a t)$$

$$= \sum_r (p_{rf} \delta_f q_r - p_{ra} \delta_a q_r) - (H_f \delta_f t - H_a \delta_a t). \qquad . \qquad . \quad (53)$$

Let the virtual path now be subjected to no restrictions concerning its energy, but suppose it to be co-terminous in space *and time* with the real path.   Eq. (53) then reduces to

$$\boxed{\delta \int L dt = 0}. \qquad . \qquad . \qquad . \quad (54)$$

This equation is known as "Hamilton's principle."   We may enunciate it as follows: *a system moves from one configuration to another in such a way that the variation of the integral $\int L dt$ between the path taken and a neighbouring virtual path, co-terminous in space and time with the actual path, is zero.*   In other words, $\int L dt$ is stationary.

A variational principle of this kind can be treated by means of the calculus of variations which exhibits the equivalence between such a principle and a set of differential equations, namely the Euler conditions.†   The method is applicable if we

† The proof of these equations is given in the Appendix.

are given a function f of three variables, and it is required to determine a curve $x = x(t)$, such that the variation between the integral $\int f\left(x, \dfrac{dx}{dt}, t\right) dt$ taken over the curve and the integral taken over an adjacent curve is zero. The desired curve is supposed to pass through two fixed points (x_1, t_1), (x_2, t_2); the integral is to be taken between these two points. The necessary and sufficient condition for the curve to possess this variational property is that it satisfies the differential equation

$$\frac{d}{dt}\left(\frac{\partial f}{\partial\left(\dfrac{dx}{dt}\right)}\right) - \frac{\partial f}{\partial x} = 0. \quad . \quad . \quad . \quad (55)$$

Consider now the case where there are several *dependent* variables $x, y, \ldots$, and f is a function of $x, y, \ldots, \dfrac{dx}{dt}, \dfrac{dy}{dt}, \ldots, t$. Again, we are given two fixed points $(x_1, y_1, \ldots, t_1)$, $(x_2, y_2, \ldots, t_2)$, and we must subject the integral $\int f\, dt$, taken between the fixed points, to the variational condition. The equations of the curve will then read

$$\left.\begin{aligned}
\frac{d}{dt}\left(\frac{\partial f}{\partial\left(\dfrac{dx}{dt}\right)}\right) - \frac{\partial f}{\partial x} = 0 \\[2ex]
\frac{d}{dt}\left(\frac{\partial f}{\partial\left(\dfrac{dy}{dt}\right)}\right) - \frac{\partial f}{\partial y} = 0 \\[2ex]
. \quad . \quad . \quad . \quad . \quad . \quad .
\end{aligned}\right\} \quad . \quad . \quad . \quad . \quad (56)$$

If we have several *independent* variables $s, t, \ldots$ and several *dependent* variables $x, y, \ldots$, the problem is slightly more complicated. Starting from a function

$$f\left(x, y, \ldots, \frac{\partial x}{\partial s}, \frac{\partial x}{\partial t}, \ldots, \frac{\partial y}{\partial s}, \frac{\partial y}{\partial t}, \ldots, s, t, \ldots\right),$$

our aim is to find functions

$$x = x(s, t, \ldots)$$
$$y = y(s, t, \ldots)$$
$$. \quad . \quad . \quad . \quad . \quad . \quad ,$$

for which

$$\int f\left(x, y, \ldots, \frac{\partial x}{\partial s}, \frac{\partial x}{\partial t}, \ldots, \frac{\partial y}{\partial s}, \frac{\partial y}{\partial t}, \ldots, s, t, \ldots\right) ds dt \ldots$$

is stationary with respect to small variations of $x, y, \ldots$. This integral is to be taken over a fixed region of the independent variables; the values of the dependent variables at the boundaries of this region are regarded as fixed. The differential equations for this problem, which were worked out by Lagrange, are

$$\left.\begin{aligned}
\frac{\partial}{\partial s}\left(\frac{\partial f}{\partial\left(\frac{\partial x}{\partial s}\right)}\right) + \frac{\partial}{\partial t}\left(\frac{\partial f}{\partial\left(\frac{\partial x}{\partial t}\right)}\right) + \ldots - \frac{\partial f}{\partial x} = 0 \\[2em]
\frac{\partial}{\partial s}\left(\frac{\partial f}{\partial\left(\frac{\partial y}{\partial s}\right)}\right) + \frac{\partial}{\partial t}\left(\frac{\partial f}{\partial\left(\frac{\partial y}{\partial t}\right)}\right) + \ldots - \frac{\partial f}{\partial y} = 0 \\[2em]
\cdots \cdots \cdots \cdots \cdots \cdots \cdots
\end{aligned}\right\} \quad \cdot \quad (57)$$

Reverting now to Hamilton's principle, we recognize that *the Euler conditions* (56) *for the integral* $\int L dt$ *to be stationary are precisely Lagrange's equations of motion* (33) *for the system.* We are accordingly led to a new approach to this variational principle, by which its direct connexion with the equations of motion of the system is disclosed. Alternatively, Hamilton's principle can be proved, using a rectangular co-ordinate system; it is then equally valid in a generalized co-ordinate system and the Lagrange equations of motion can be deduced therefrom. One of the most salient characteristics of variational principles of the type (52) or (54) is here revealed, namely that they express the pertinent laws without involving a co-ordinate system. Hence, if a set of differential equations related to a particular co-ordinate system can be identified with a variational principle, one may, in many cases, apply the Euler-Lagrange conditions, which give the equations in a form independent of the choice of co-ordinates.

The method discussed in the last paragraph provides us with a second, comparatively simple, proof of the invariance of Lagrange's equations of motion. However, any such invariance property can be derived directly, as has been shown in §5.

And it is worth mentioning that that proof is perhaps the less circuitous and does not introduce the calculus of variations which may be regarded as irrelevant if we are interested in the invariance of the equations alone.

Thus far, equations (23), (50) and (53), which represent respectively the variations of the action integral (Lagrange) and the integral $\int L dt$ (Hamilton) between two paths whose endpoints do not coincide, have only been employed as a preparatory step in proceeding to the case where the paths are co-terminous, i.e., to prove the principle of least action and Hamilton's principle. It must be admitted that this approach has not resulted in any new equations; it has merely confirmed those already established by other means. The solution of dynamical problems is therefore not greatly facilitated. The admirable insight of Hamilton into the implications of the general equations (23), (50) and (53) as they stand, must be appreciated as a fundamental advance in the theory of mechanics. He considered the action as a function of the initial point of the path taken, the final point and the energy,—these $(2n + 1)$ quantities are necessary and sufficient to determine uniquely† the path and hence the action.[15, 16] The action integral so expressed is known as "Hamilton's characteristic function" S,‡ and the formulae (23) or (50) for its differential were called by him the "law of varying action." We may once more repeat Hamilton's own words, as he enjoyed an international reputation, not only as a scientist, but also for his elegant and forceful style.

Yet from not having formed the conception of the action as a *function* of this kind, the consequences that have been here deduced from the formula (A.)§ for the variation of that definite integral, appear to have escaped the notice of Lagrange. . . . For although Lagrange and others, in treating of the motion of a system have shown that the variation of this definite integral vanishes when the extreme co-ordinates and the constant H are given, they appear to have deduced from this result only the well known law of *least action*. . . . But when this well known law of least, or as it might better be called, of *stationary action*, is applied to the determination of the actual motion of a system, it serves only to form, by the rules of the calculus of variations, the

---

† It is, however, possible that two or three paths satisfying the required conditions exist, e.g., a particle may travel between two fixed points on a Kepler orbit taking the shorter or the longer route. The function is then multi-valued.

‡ i.e., $S(q_{xa}, q_{xf}, E) = \left(\int_{q_{xa}}^{q_{xf}} \sum_s p_s \, dq_s \right)_{H=E}$.

§ Corresponding to eq. (23).

differential equations of motion of the second order, which can always be otherwise found. It seems, therefore, to be with reason that Lagrange, Laplace, and Poisson have spoken lightly of the utility of this principle in the present state of dynamics. A different estimate, perhaps, will be formed of that other principle which has been introduced in the present paper, under the name of the *law of varying action*, in which we pass from an actual motion to another motion dynamically possible, by varying the extreme positions of the system, and (in general) the quantity *H*, and which serves to express, by means of a single function, not the mere differential equations of motion, but their intermediate and their final integrals.

From the law of varying action (50), the derivatives of *S* with respect to the initial and final co-ordinates can immediately be written down. They assume the form

$$(a) \ \frac{\partial S}{\partial q_{rf}} = p_{rf}, \quad (b) \ \frac{\partial S}{\partial q_{ra}} = -p_{ra}. \qquad . \qquad . \qquad (58)$$

Since the energy has to be specified, we have

$$H(q_r, p_r) = E.$$

Substituting from (58) we arrive at the two partial differential equations which Hamilton put forward as an adequate tool for determining *S*, namely

$$\left. \begin{array}{l} (a) \ H\left(q_{rf}, \dfrac{\partial S}{\partial q_{rf}}\right) = E \\[3ex] (b) \ H\left(q_{ra}, \dfrac{\partial S}{\partial q_{ra}}\right) = E \end{array} \right\} \qquad . \qquad . \qquad . \qquad (59)$$

How did Hamilton utilize these equations to solve dynamical problems? Eqs. (59) have to be integrated to find *S*,† which is treated as a function of the fixed initial points q_{ra}, the variable end-points q_{rf} and the energy *E*. In virtue of (58), one can then differentiate *S* with respect to any $n - 1$ of the fixed initial points q_{ra} and set these derivatives respectively equal to constants p_{ra}, thereby obtaining $(n - 1)$ equations in the variables q_{rf}. These equations contain $(2n - 1)$ constants

---

† It is, indeed, by no means obvious that any solution of (59) will furnish the action function *S*. At present, our sole objective is a sketchy description of Hamilton's own approach. The further development will be studied rigorously in the following pages.

(q_{ra}, p_{ra}, E, where r runs from 1 to $n - 1$) and define the paths of the system.   The remaining equation of (58b) is automatically fulfilled according to (59b).   The relations (58a) finally enable us to find the momenta at any point of the path.

Hamilton first "verified" that the procedure just outlined leads, in the case of rectangular co-ordinates, to Newton's laws of motion.   He then showed, by direct transformation, that in generalized co-ordinates eq. (50) corresponds to eq. (23). In this manner, he extended the method of the characteristic function to arbitrary co-ordinate systems, and next verified that it was equivalent to Lagrange's equations, which are thus derived from a new angle.   Since the whole treatment has more recently been improved and widened, we shall not elaborate his verifications, but they are, in essence, the same as those given later in the discussion of the completed theory.

In dealing with Hamilton's characteristic function, which involves the principle of varying action, we have to know the energy of the system.   (As mentioned previously, S depends on the parameter E.)   When we adopt as our starting-point eq. (53), which relates to Hamilton's principle and not to the principle of least action, E is eliminated and, moreover, the time t is brought into consideration.   In place of S, let W be equal to $\int L dt$, which is regarded as a function of the fixed initial points, the variable end-points and the time t over which the integral is taken. W is called "Hamilton's principal function."   Eq. (58) is now substituted by (cf. eq. (53))

$$(a)\ \frac{\partial W}{\partial q_{rf}} = p_{rf}; \quad (b)\ \frac{\partial W}{\partial q_{ra}} = -p_{ra}; \quad (c)\ \frac{\partial W}{\partial t} = -E, \quad (60)$$

and (59) therefore becomes

$$\left. \begin{aligned} (a)&\ H\left(q_{rf}, \frac{\partial W}{\partial q_{rf}}\right) + \frac{\partial W}{\partial t} = 0 \\[2mm] (b)&\ H\left(q_{ra}, \frac{\partial W}{\partial q_{ra}}\right) + \frac{\partial W}{\partial t} = 0\dagger \end{aligned} \right\} \quad . \quad . \quad . \quad (61)$$

Again, eqs. (61) are to be integrated to find W, and (60b) then provides n equations connecting the co-ordinates of the particle and the time t.   The equations contain $2n$ arbitrary constants

---

† Since H is a quadratic function of the p's, it is unnecessary to write $-\dfrac{\partial W}{\partial q_{ra}}$.

q_{ra}, p_{ra}, and suffice to describe the motion. The momenta can be obtained from (60a), and eq. (60c) is immediately satisfied owing to (61). Hamilton did not verify that this later method produces correct results when applied to dynamical problems, as he had done in full for the characteristic function in his earlier paper. It appears as if he took it for granted that the integration of (61) gives uniquely the desired function W, but this is an assumption which demands proof. On this evidence, one is probably justified in concluding that, in dealing with his characteristic function, the "verification" was not meant as a proof in the strict sense of the word, but rather as a confirmation of the legitimacy of his procedure.

While Hamilton must be given all credit for the introduction of this new method which involves the search for only one function (S or W) and which serves to derive the equations of motion in non-differential form, the full scope and generality of this theory was not realized before Jacobi (1804–1851). The constants which occurred in the function S or W investigated by Hamilton had to be the initial co-ordinates (and, in S, the energy E as well). Jacobi,[17] on the other hand, demonstrated that this restriction was unwarranted, and that, instead of (59a and b) or (61a and b), *one* differential equation alone was needed, namely

$$\boxed{H\left(q_r, \frac{\partial S}{\partial q_r}\right) = E}\,, \qquad . \qquad . \qquad . \qquad (62)$$

or

$$\boxed{H\left(q_r, \frac{\partial W}{\partial q_r}\right) + \frac{\partial W}{\partial t} = 0}\,. \qquad . \qquad . \qquad (63)$$

Equations (62) and (63), being essentially identical with (59a) and (61a) respectively, are the two alternative versions of "Hamilton's partial differential equation" or the "Hamilton-Jacobi equation." Jacobi concentrated on eq. (63) and proved that *any* complete integral of this equation, i.e., an integral containing as many arbitrary constants as there are independent variables (in this case $n + 1$), may be used. One of the constants is simply combined with W by addition; denoting the others by α_r ($r = 1, 2, \ldots, n$), we shall show below that the system moves in such a way that the derivatives of W with respect to the α's would remain constant with time, and the equations of motion of the system (in non-differential form) accordingly read

$$\frac{\partial W}{\partial \alpha_r} = \beta_r \quad , \qquad . \qquad . \qquad . \qquad (64)$$

where β_r are n further arbitrary constants. Eqs. (64) are sometimes known as Jacobi's theorem. These equations thus contain $2n$ arbitrary constants α_r and β_r. The momenta at any point of the path can be calculated from the formulae

$$\frac{\partial W}{\partial q_r} = p_r \quad . \qquad . \qquad . \qquad . \qquad (65)$$

In a similar way, dynamical problems can be solved with the aid of eq. (62). A complete integral will now require n arbitrary constants as well as E. Except for one which is additive, they may be symbolized by α_r $(r = 1, 2, \ldots, n - 1)$. Instead of (64), we may write

$$\frac{\partial S}{\partial \alpha_r} = \beta_r \quad , \qquad . \qquad . \qquad . \qquad (66a)$$

which provide the $(n - 1)$ equations of the orbit (not including the time) and contain $(2n - 1)$ constants α_r, β_r, E. From

$$\frac{\partial S}{\partial q_r} = p_r \quad , \qquad . \qquad . \qquad . \qquad (67)$$

the momenta can be found as before.

If the integrated equations of motion of a system are desired —given the co-ordinates and momenta at a certain time t—the n eqs. (65) can be solved for the n constants α_r. By direct differentiation we obtain the β_r, thus determining the motion fully. The stipulation that (65) furnishes one, and only one, value for each α, is to be implied in the statement that W contains n arbitrary constants (besides an additive one). Thus the co-ordinates and momenta of the system at time t serve to determine uniquely the equations of motion; this condition must, as is well-known, hold in any correct theory of dynamics. In a like manner, when one starts out from (62) in place of (63), the $(n - 1)$ α's can be computed from any $n - 1$ of the eqs. (67).

The nth of the eqs. (67) is then automatically satisfied, owing to the energy condition and eq. (62). As in the former case, S must be such that the constants α are uniquely defined in this way.

Jacobi's theorem can now be proved without great effort. Suppose the system is moving in conformity with (64), and let us *define* the p_r's by (65). Then eq. (64) may be stated as

$$\frac{d}{dt}\left(\frac{\partial W}{\partial \alpha_r}\right) = 0,$$

or
$$\frac{\partial}{\partial t}\left(\frac{\partial W}{\partial \alpha_r}\right) + \sum_s \frac{\partial^2 W}{\partial q_s\,\partial \alpha_r}\,\dot{q}_s = 0. \qquad . \qquad . \qquad (68)$$

Further, differentiating (63) with respect to α_r gives

$$\frac{\partial}{\partial \alpha_r}\left(\frac{\partial W}{\partial t}\right) + \sum_s \frac{\partial^2 W}{\partial \alpha_r\,\partial q_s}\cdot\frac{\partial H}{\partial\left(\dfrac{\partial W}{\partial q_s}\right)} = 0,$$

i.e.,
$$\frac{\partial}{\partial \alpha_r}\left(\frac{\partial W}{\partial t}\right) + \sum_s \frac{\partial^2 W}{\partial \alpha_r\,\partial q_s}\cdot\frac{\partial H}{\partial p_s} = 0. \qquad . \qquad (69)$$

With regard to the n equations (68) for the n quantities $\dot{q}_r$, the determinant $\left|\dfrac{\partial^2 W}{\partial \alpha_r\,\partial q_s}\right|$ is non-zero, as its disappearance would mean the vanishing of the Jacobian $\dfrac{\partial(p_1, p_2, \ldots, p_n)}{\partial(\alpha_1, \alpha_2, \ldots, \alpha_n)}$; this in turn implies that the p's cannot be varied independently and so contradicts the condition imposed upon W. Eq. (68) therefore has a unique solution which is at once observed, on comparison with (69), to be

$$\dot{q}_r = \frac{\partial H}{\partial p_r}.$$

By differentiation of (63) with respect to q_r, we find

$$\frac{\partial H}{\partial q_r} + \sum_s \frac{\partial H}{\partial\left(\dfrac{\partial W}{\partial q_s}\right)}\cdot\frac{\partial^2 W}{\partial q_r\,\partial q_s} + \frac{\partial^2 W}{\partial q_r\,\partial t} = 0, \qquad . \qquad (70)$$

i.e.,
$$\frac{\partial H}{\partial q_r} + \sum_s \frac{\partial H}{\partial p_s}\frac{\partial^2 W}{\partial q_r\,\partial q_s} + \frac{\partial^2 W}{\partial q_r\,\partial t} = 0$$

and therefore
$$\frac{\partial H}{\partial q_r} + \sum_s \dot{q}_s \frac{\partial^2 W}{\partial q_r \, \partial q_s} + \frac{\partial^2 W}{\partial q_r \, \partial t} = 0,$$

or
$$\frac{\partial H}{\partial q_r} + \frac{d}{dt}\left(\frac{\partial W}{\partial q_r}\right) = 0,$$

i.e.,
$$\dot{p}_r = -\frac{\partial H}{\partial q_r}.$$

Hence the Hamilton canonical equations are fulfilled, and the equations (64) represent dynamically possible states of motion. Conversely, every state of motion which conforms to the ordinary laws of mechanics will satisfy (64), because—as already pointed out—suitable α's and β's can be calculated, given any co-ordinates and momenta at time t.

The proof of the other form of Jacobi's theorem, namely that dealing with (62) instead of (63), can be evolved by a very similar procedure: the equation

$$\sum_s \frac{\partial^2 S}{\partial q_s \, \partial \alpha_r} \dot{q}_r = 0 \qquad . \qquad . \qquad . \qquad (71)$$

must be substituted for (68), and (69) is to be replaced by

$$\sum_s \frac{\partial^2 S}{\partial \alpha_r \, \partial q_s} \frac{\partial H}{\partial p_s} = 0. \qquad . \qquad . \qquad (72)$$

In this case, (71) consists of $(n-1)$ homogeneous linear equations for the q_r's, and the disappearing of the determinant $\dfrac{\partial^2 S}{\partial q_s \, \partial \alpha_r}$ (r, s running from 1 to $n-1$) entails that $n-1$ of the n momenta cannot be varied independently, and that the imposed conditions are therefore once more violated. Consequently, (71) specifies uniquely the ratios between the $\dot{q}_r$'s, which can be seen from (72) to be the same as the ratios between the $\dfrac{\partial H}{\partial p_r}$'s. The energy condition, together with (72), enables us to write

$$\dot{q}_r = \frac{\partial H}{\partial p_r}.$$

The second of the Hamilton equations (48) can be developed exactly as in the previous deduction; we must omit all terms that contain partial derivatives with respect to time. Accordingly, the application of the Hamilton-Jacobi method always leads to permissible motions, and the converse follows as before.†

The relation between the position of the particle in its orbit and the time may also be worked out if required. For, differentiating (62) with respect to E, we get

$$\sum_r \frac{\partial H}{\partial \left(\frac{\partial S}{\partial q_r}\right)} \frac{\partial}{\partial E} \left(\frac{\partial S}{\partial q_r}\right) = 1,$$

i.e.,

$$\sum_r \frac{\partial H}{\partial p_r} \frac{\partial}{\partial q_r} \left(\frac{\partial S}{\partial E}\right) = 1$$

and thus

$$\sum_r \frac{\partial}{\partial q_r} \left(\frac{\partial S}{\partial E}\right) \dot{q}_r = 1,$$

i.e.,

$$\frac{d}{dt} \left(\frac{\partial S}{\partial E}\right) = 1,$$

or

$$\frac{\partial S}{\partial E} = \beta_0 + t, \quad . \quad . \quad (66b)$$

which furnishes the desired equation.

Although Jacobi's function S, as opposed to the original action function of Hamilton, is not the action along the path between two points, it is nevertheless evident from (67) that the action integral between any two points of an orbit is given by the difference between the respective values of S. Needless to say, the appropriate values of the constants α_r must be used.

The integral $\int L dt$ between two points on a path is similarly equal to the difference of the respective values of W, Hamilton's principal function (as modified by Jacobi). This is directly deducible, because

$$\text{change in } W = \sum_r \int \frac{\partial W}{\partial q_r} dq_r + \int \frac{\partial W}{\partial t} dt$$

$$= \sum_r \int p_r dq_r - \int H dt, \text{ from (65) and (63)},$$

$$= \int \left(\sum_r p_r \dot{q}_r - H\right) dt$$

$$= \int L dt, \text{ from (47)}.$$

† We shall in §10 discuss how the Hamilton-Jacobi equation can be integrated in a frequently occurring class of cases.

Of the two formulations of Jacobi's theorem, that involving Hamilton's characteristic function S is the more convenient for quantum theory owing to its close relation to $\int p_r dq_r$. On the other hand, eq. (63) is more general, being applicable to systems in which the Hamiltonian contains the time; apart from this, it has the advantages which were mentioned earlier in this section in connexion with the introduction of (61).

As a point of historical interest, it may be recorded that Jacobi himself treated only that form of the theory based upon the function W, and restricted himself to rectangular co-ordinates. While appreciating without reservation Hamilton's achievement, Jacobi yet seemed critical of him for not having arrived directly at the general form of the theory.

> I therefore do not know why Hamilton, in order to be able to indicate the general integrals of the above differential equations, requires the introduction of a function S† of $6n + 1$ variables, namely the $3n$ quantities x_i, y_i, z_i, the $3n$ quantities a_i, b_i, c_i, and the quantity t, which function satisfies at the same time *both* partial differential equations of the first order
>
> $$\frac{\partial S}{\partial t} + \frac{1}{2} \sum \frac{1}{m_i} \left[\left(\frac{\partial S}{\partial x_i} \right)^2 + \left(\frac{\partial S}{\partial y_i} \right)^2 + \left(\frac{\partial S}{\partial z_i} \right)^2 \right] = U,‡$$
>
> $$\frac{\partial S}{\partial t} + \frac{1}{2} \sum \frac{1}{m_i} \left[\left(\frac{\partial S}{\partial a_i} \right)^2 + \left(\frac{\partial S}{\partial b_i} \right)^2 + \left(\frac{\partial S}{\partial c_i} \right)^2 \right] = U_0,$$
>
> while, as we have seen, it is completely sufficient to know any function of the $3n + 1$ quantities t, x_i, y_i, z_i, which satisfies the one equation
>
> $$\frac{\partial S}{\partial t} + \frac{1}{2} \sum \frac{1}{m_i} \left[\left(\frac{\partial S}{\partial x_i} \right)^2 + \left(\frac{\partial S}{\partial y_i} \right)^2 + \left(\frac{\partial S}{\partial z_i} \right)^2 \right] = U$$
>
> and contains, apart from the constant combined with S by addition, $3n$ further arbitrary constants.

Before we conclude this section, it might be apposite to give a brief survey of the manner in which Hamilton was led to the discovery of the great power and usefulness of the function S in dynamics. His starting-point was the analogy between mechanics and optics.[18] This analytical relationship was of some importance in Hamilton's time, because of the formal connexion between the wave and corpuscular theories of light which it revealed; with the advent of Schrödinger's wave mechanics, moreover, a renewed interest has been aroused in the topic.

† In our notation, W.
‡ Jacobi employs U instead of $-V$

We already know from §2 that, whenever a ray of light is reflected or refracted at a surface, the quantity $\sum \mu s$, where s is the distance travelled in a medium and μ the refractive index, satisfies our variational condition. Without framing any hypothesis as to the nature of light, Hamilton defined this quantity as the action. The variational principle can immediately be extended to the more general case of several surfaces. He then considered a fixed light source and examined the "surfaces of constant action," i.e., the surfaces over which the action, taken along the rays emitted from the fixed initial source, is constant. By means of the principle of stationary

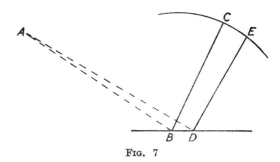

Fig. 7

action, he proved in a simple manner that the rays cut these surfaces orthogonally. Let ABC and ADE (Fig. 7) represent two adjacent rays from the source A to a surface of constant action CE, and let them cut the last reflecting or refracting surface along their paths at B and D respectively. According to the principle of stationary action,

total action along ABC = total action along ADC.

Also, CE being a surface of constant action,

total action along ABC = total action along ADE,

so   total action along ADC = total action along ADE,

whence $\qquad\qquad DC = DE$,

which gives naturally the required result.

We are now able to derive the principle of varying action; for, from the theorem just developed, if S is the action function,

$$\text{grad } S = \frac{\partial S}{\partial s}$$

$$= \mu,$$

from which $\qquad dS = \mu(l\,dx + m\,dy + n\,dz),\qquad .\qquad .\qquad (73)$

where l, m and n are the direction-cosines of the ray with respect to the co-ordinate axes. The derivatives of the action function thus become

$$\frac{\partial S}{\partial x} = \mu l; \quad \frac{\partial S}{\partial y} = \mu m; \quad \frac{\partial S}{\partial z} = \mu n. \qquad . \qquad (74)$$

As the next step, S is regarded as a function of the co-ordinates a, b, c of the source as well as of the co-ordinates of the end-point, and our function satisfies the three further equations

$$\frac{\partial S}{\partial a} = -\mu_0 l_0; \quad \frac{\partial S}{\partial b} = -\mu_0 m_0; \quad \frac{\partial S}{\partial c} = -\mu_0 n_0. \quad (75)$$

All these theorems and formulae can at once be generalized to non-homogeneous media by increasing the number of surfaces *ad infinitum*. The function S is then defined as

$$S = \int \mu ds, \qquad . \qquad . \qquad . \qquad . \qquad (76)$$

and the principle of least action adopts the form

$$\delta \int \mu ds = 0. \qquad . \qquad . \qquad . \qquad (77)$$

Hamilton recognized that, owing to (74), (75) and the equation $l^2 + m^2 + n^2 = 1$, S must comply with the partial differential equations

$$\left. \begin{array}{l} (a) \quad \left(\dfrac{\partial S}{\partial x}\right)^2 + \left(\dfrac{\partial S}{\partial y}\right)^2 + \left(\dfrac{\partial S}{\partial z}\right)^2 = \mu^2 \\[4mm] (b) \quad \left(\dfrac{\partial S}{\partial a}\right)^2 + \left(\dfrac{\partial S}{\partial b}\right)^2 + \left(\dfrac{\partial S}{\partial c}\right)^2 = \mu_0^2 \end{array} \right\} . \quad . \quad (78)$$

The method of attack was to integrate eqs. (78) in order to determine S. Any two of the three formulae (75) would then furnish the equations of the ray, containing five arbitrary constants, namely a, b, c and two out of l_0, m_0, n_0. As is to be expected, the third equation of (75) would be obeyed because of (78b) and the relation $l^2 + m^2 + n^2 = 1$. By simple differentiation of S with respect to the co-ordinates, we could finally obtain from (74) the direction-cosines at any point on the ray. Just as in his later work on dynamics, Hamilton did not demonstrate rigorously that S, given by (78), was necessarily the correct function.

So far, the whole treatment has been independent of any theories regarding the actual constitution of light. When we turn to the two hypotheses current in Hamilton's day, a comparison with eq. (15) discloses that, in the case of the corpuscular theory, the potential energy of the light particles in any medium is such that their velocity is directly proportional to the refractive index. The reader will recall that this result had already been attained, though rather crudely, by Maupertuis.

In the undulatory theory, however, the velocity is inversely proportional to the refractive index, and the quantity S in (76) is the same as the time taken for the light to travel between two points. The basic and remarkable conclusion thus exhibited enables us to identify the constructs in Hamilton's theory with familiar concepts. The surfaces of constant action reveal themselves as no other than the wave surfaces; the orthogonality to these surfaces of the rays is a fundamental factor in wave optics; and lastly the principle of least action becomes Fermat's principle of least time in its variational form, which has been proved quite generally in §2 for wave motion.

Hamilton later extended this optical method to dynamics, as has been dealt with in fair detail earlier in this paragraph. A brief reflection suffices to elucidate how the inner mathematical connexion between mechanics and optics lies, in essence, in the analogy of the principle of least action to Fermat's principle. It is clear that, in the simple case of a particle moving in a potential field, the principle of least action predicts the same path as Fermat's principle would for a light ray in a heterogeneous medium, provided the refractive index be defined by

$\mu \propto$ momentum,

i.e., $\quad \mu \propto \sqrt{E - V}$ $\qquad$ (in classical theory)

$\mu \propto \sqrt{\dfrac{1}{c^2}(E - V + m_0 c^2)^2 - m_0{}^2 c^2}$ (in relativity theory)

$\left.\begin{array}{}\\ \\ \\ \\ \end{array}\right\}$. . (79)

The last formula is derived from the well-known equation

$$\frac{1}{c^2}(E - V + m_0 c^2)^2 = p_x{}^2 + p_y{}^2 + p_z{}^2 + m_0{}^2 c^2.$$

It ought to be emphasized that it is the principle of least action and not Hamilton's principle which best expresses the relation of mechanics and optics, because in the latter there exists no variable corresponding to the energy of a particle, and the time taken for the ray to pass between two points is a fixed quantity. In stating formally the connexion between these two branches of physics, the energy of the particles in the dynamical system has therefore first to be specified, that is, we compare the paths of particles with a certain fixed energy to those of light rays.

The foregoing analogy has been propounded in a slightly different and perhaps more illuminating manner by Schrödinger, who adopted it as the foundation for his wave mechanics.[19] He regarded the configuration space as non-Euclidean, with the line element given by

$$ds^2 = 2\bar{T}(q_r, \dot{q}_r)\, dt^2, \qquad . \qquad . \qquad (80)$$

where $\bar{T}$ is the kinetic energy expressed as a function of the generalized *velocities* and co-ordinates. As $\dot{q}_r dt = dq_r$, $\bar{T}dt^2$ is a quadratic form in the dq_r's. The whole idea is, of course, no more than a mathematical construction, but this should not prevent us from invoking it in order to obtain an interpretation of the usual terms of vector analysis—perpendicularity, gradient, etc. For these terms can be thought of as if they applied in an ordinary three-dimensional Euclidean space, but it must be borne in mind that, from a mathematical standpoint, the space is n-dimensional, non-Euclidean and has co-ordinates with the line element defined by (80). $\bar{T}$ being a scalar, $\partial\bar{T}/\partial\dot{q}_r$, viz., p_r is a co-variant vector. It follows that the tensor formed by the coefficients of $p_i p_j$ in the quantity $T(q_r, p_r)$, where T is the kinetic energy of the system, expressed as a function of the generalized *momenta* and co-ordinates, is contra-variant.

The artifice of this non-Euclidean metric gives the Hamilton-Jacobi equation a neat, simple form. The arguments set forth in the previous paragraph yield the result that $T(q_r, \partial S/\partial q_r)$ is a scalar, which, from (80) and the definition of p_r, is seen without difficulty to be $\frac{1}{2}(\text{grad } S)^2$. Eq. (62) now translates into

$$(\text{grad } S)^2 = 2(E - V). \qquad . \qquad . \qquad (81a)$$

This formula is, needless to say, only valid in classical mechanics. In relativity theory, we restrict ourselves in any case to a single particle moving in a potential field and, for this particular problem, we do not require a non-Euclidean space, but may immediately write the Hamilton-Jacobi equation as

$$\left(\frac{\partial S}{\partial x}\right)^2 + \left(\frac{\partial S}{\partial y}\right)^2 + \left(\frac{\partial S}{\partial z}\right)^2 + m_0^2 c^2 = \frac{1}{c^2}\left(E - V + m_0 c^2\right)^2$$

or $\qquad (\text{grad } S)^2 = \dfrac{1}{c^2}\left(E - V + m_0 c^2\right)^2 - m_0^2 c^2. \qquad . \qquad (81b)$

In order to find the solution of (81*a*) or (81*b*), the following construction may be employed (Fig. 8). Take as a starting-point any $(n-1)$-dimensional surface over which the action function is constant and equal to S_0. Draw the normal at each point to the surface and, along these normals, mark off, in either one of the two directions, distances $ds = \dfrac{dS}{\sqrt{2(E-V)}}$ (or $\dfrac{dS}{\sqrt{\dfrac{1}{c^2}(E-V+m_0c^2)^2 - m_0{}^2c^2}}$ in relativity theory). The surface over which the action function has the value $S_0 + dS$ is

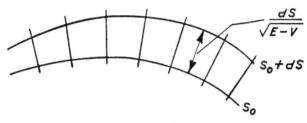

FIG. 8

then the surface passing through these points. Similarly, the construction can be carried out over the whole space. Eq. (67) may be stated, in the non-Euclidean space,

$$\operatorname{grad} S = \mathbf{p},$$

whence we conclude that the velocity of the system is perpendicular to the surfaces of constant action.

The passage from this aspect of the Hamilton-Jacobi theory to optics is now possible without difficulty. Thus, let a train of waves move with velocity proportional to $\dfrac{1}{\sqrt{E-V}}$ (or $\dfrac{1}{\sqrt{\dfrac{1}{c^2}(E-V+m_0c^2)^2 - m_0{}^2c^2}}$ in relativity theory), i.e., as if the refractive index of the medium were proportional to $\sqrt{E-V}$ (or $\sqrt{\dfrac{1}{c^2}(E-V+m_0c^2)^2 - m_0{}^2c^2}$ in relativistic mechanics). Suppose further that one of the wave surfaces coincides with our initial surface of constant action. Then it can at once be seen that, as the wave progresses, the wave surface will continue to coincide with one of the surfaces of constant action. The rays will be orthogonal to these surfaces, and will therefore be identical with the paths of the dynamical system.

Recapitulating, one may maintain without hesitation that the formal analogy of mechanics to optics represents the historical basis of one of the most potent devices at our disposal for the mathematical analysis of dynamical systems. By an irony, it was this very relationship that inspired de Broglie, and in particular Schrödinger, to establish the theory of wave mechanics which led finally to the overthrow of Newtonian mechanics as an ultimate description of the physical facts of nature.

§ 7

Contact Transformations and Hamilton's Canonical Equations

HAMILTON'S canonical equations of motion are, as has been explained previously, invariant under transformations of the form

$$Q_r = Q_r(q_1, q_2, \ldots, q_n, t).$$

The equations remain, however, invariant under a much wider class of transformations. For, let

$$Q_r = Q_r(q_1, q_2, \ldots, q_n, p_1, p_2, \ldots, p_n, t)$$
$$P_r = P_r(q_1, q_2, \ldots, q_n, p_1, p_2, \ldots, p_n, t).$$

It is required to determine the conditions under which Hamilton's equations transform invariantly. Define the function

$$\sum_r p_r \dot{q}_r - H(q, p),$$

where the $\dot{q}$'s and p's are regarded as *independent of each other*. Inspection shows that the Euler conditions (56) for the time integral between two points of this function to be stationary are

$$\dot{p}_r + \frac{\partial H}{\partial q_r} = 0,$$

$$\dot{q}_r - \frac{\partial H}{\partial p_r} = 0,$$

which are precisely Hamilton's equations. It must be emphasized that, though $\sum_r p_r \dot{q}_r - H = L$, this variational condition is nevertheless more general than Hamilton's principle, because the p's and q's vary independently.

Assume now that we have some function $\bar{H}$ such that

$$\sum_r P_r \dot{Q}_r - \bar{H}(Q_r, P_r) = \sum_r p_r \dot{q}_r - H(q_r, p_r) - \frac{dF}{dt}, \qquad (82)$$

where F be any function of the Q_r's, the q_r's and t, or therefore of the q_r's, p_r's and t. The transformed equations will then read

$$
\left.
\begin{array}{ll}
(a) & \dfrac{\partial \bar{H}}{\partial P_r} = \dot{Q}_r \\[3mm]
(b) & \dfrac{\partial \bar{H}}{\partial Q_r} = -\dot{P}_r
\end{array}
\right\}, \qquad . \qquad . \qquad . \quad (83)
$$

since the stationary nature of $\int (\sum_r p_r \dot{q}_r - H) dt$ depends in no way on the co-ordinate system, and $\int \dfrac{dF}{dt}\, dt$, being simply the difference between the values of F for the extreme points of the range of integration, is once more independent of the co-ordinates. We recognize here another instance of the applicability of a variational principle to deduce the invariance property of a set of equations.

Alternatively, one can prove the invariance of Hamilton's equations independently of the calculus of variations. To attain this objective, we need merely draw the reader's attention to the proof of eq. (38), where it was rigorously shown that any set of equations of the Euler-Lagrange form is invariant under the type of transformation discussed. The circumstance that we encounter, in this case, p's as well as q's, is irrelevant: the p's are not related to the q's and may, if we wish to do so, be designated by $q_{n-1}, q_{n-2}, \ldots, q_{2n}$. The insertion of dF/dt on the right side of (82) is justified, since it will now be demonstrated that the Euler-Lagrange equations are satisfied identically if the function to which they are applied is a total derivative. Developing the derivatives in the usual way,

$$
\frac{d}{dt}\left\{ \frac{\partial \left(\dfrac{dF}{dt} \right)}{\partial \dot{q}_r} \right\} - \frac{\partial \left(\dfrac{dF}{dt} \right)}{\partial q_r}
$$

$$
= \frac{d}{dt}\left\{ \frac{\partial}{\partial \dot{q}_r} \left(\sum_s \frac{\partial F}{\partial q_s}\dot{q}_s + \frac{\partial F}{\partial t} \right) \right\} - \frac{\partial}{\partial q_r}\left(\frac{\partial F}{\partial t} \right)
$$

$$
= \frac{d}{dt}\left(\frac{\partial F}{\partial q_r} \right) - \frac{\partial}{\partial q_r}\left(\frac{dF}{dt} \right)
$$

$$
= \sum_s \frac{\partial^2 F}{\partial q_s\, \partial q_r}\dot{q}_s + \frac{\partial^2 F}{\partial t\, \partial q_r} - \sum_s \frac{\partial^2 F}{\partial q_r\, \partial q_s}\dot{q}_s - \frac{\partial^2 F}{\partial q_r\, \partial t}
$$

$$
= 0.
$$

Eq. (82) holds if and only if

$$
\left.
\begin{array}{ll}
(a) & \sum_r P_r dQ_r = \sum_r p_r dq_r - dF \\[2mm]
(b) & \bar{H} = H + \dfrac{\partial F}{\partial t}
\end{array}
\right\} , \qquad . \qquad (84)
$$

where the dF in (a) refers to the co-ordinates alone and not to the time. Eqs. (84) are sufficient to ensure that (48) transform to (83) in the new co-ordinate system. Co-ordinate transformations of the form (84a) are known as "contact transformations." The origin of this term is not connected with dynamics and is consequently not our concern.

We now wish to express the condition for a contact transformation without involving differentials. There may exist one or more equations connecting the Q's and the q's only—in the transformations of §5 there were n such relations. Suppose we have m equations

$$
f_\alpha(q_r, Q_r) = 0 \quad (\alpha = 1, 2, \ldots, m). \qquad . \qquad . \qquad (85)
$$

On differentiation we get

$$
\sum_r \left(\frac{\partial f_\alpha}{\partial q_r} dq_r + \frac{\partial f_\alpha}{\partial Q_r} dQ_r \right) = 0. \qquad . \qquad . \qquad (86)
$$

Since eq. (84a) holds, with the q's and Q's varying in any manner consistent with (86), we may write

$$
\left.
\begin{array}{l}
p_r = \sum_\alpha \lambda_\alpha \dfrac{\partial f_\alpha}{\partial q_r} + \dfrac{\partial F}{\partial q_r} \\[4mm]
P_r = -\sum_\alpha \lambda_\alpha \dfrac{\partial f_\alpha}{\partial Q_r} - \dfrac{\partial F}{\partial Q_r}
\end{array}
\right\} , \qquad . \qquad . \qquad (87)
$$

in which the λ's are undetermined multipliers. Eqs. (85) and (87) provide us with $(2n + m)$ equations to calculate the Q's, P's and λ's, and are therefore the required conditions for a contact transformation. They were introduced by Jacobi who used them in the theory of partial differential equations. In passing, it may be pointed out that the transformations of §5 are special cases of contact transformations, as, according to (39),

$$
\sum_r P_r \dot{Q}_r = \sum_r p_r \dot{q}_r,
$$

or

$$
\sum_r P_r dQ_r = \sum_r p_r dq_r.
$$

Eq. (84) is therefore satisfied with $F = 0$.

One may also state conditions for a contact transformation, in which F is a function of variables other than the q's and Q's—in particular, where it is a function of the q's and P's. For, writing F^* equal to $F + \sum_r P_r Q_r$, eq. (82) becomes

$$-\sum_r Q_r \dot{P}_r - \bar{H}(Q_r, P_r) = \sum_r p_r \dot{q}_r - H(q_r, p_r) - \frac{dF^*}{dt}, \quad . \quad (88)$$

or, if we multiply by dt,

$$-\sum_r Q_r dP_r - \bar{H}(Q_r, P_r)dt = \sum_r p_r dq_r - H(q_r, p_r) - \sum_r \frac{\partial F^*}{\partial P_r} dP_r$$

$$-\sum_r \frac{\partial F^*}{\partial q_r} dq_r - \frac{\partial F^*}{\partial t} dt. \quad . \quad (89)$$

Thus, equating the coefficients of the differentials, we obtain as sufficient conditions for a contact transformation

$$\boxed{(a) \ p_r = \frac{\partial F^*}{\partial q_r}; \quad (b) \ Q_r = \frac{\partial F^*}{\partial P_r}; \quad (c) \ \bar{H} = H + \frac{\partial F^*}{\partial t}} \quad . \quad (90)$$

Eqs. (90) are certainly not necessary conditions, because we have not taken into account the cases where the q's and P's are related by one or more equations.

At this juncture, it is an interesting problem to find a set of generalized co-ordinates and momenta defined so that Hamilton's canonical equations of motion read

$$(a) \quad \dot{\alpha}_r = 0; \quad (b) \quad \dot{\beta}_r = 0,$$

in which α_r and β_r are the momenta and co-ordinates respectively. It is evident that our requirement will be met if $\bar{H} = 0$. When W represents the transformation function F^* for a case which complies with this suggestion, then

$$H(q_r, p_r) + \frac{\partial W(\alpha_r, q_r)}{\partial t} = 0. \quad . \quad . \quad (91)$$

By substituting β_r and α_r for Q_r and P_r, (90) must now be written as

$$(a) \quad p_r = \frac{\partial W}{\partial q_r}; \quad (b) \quad \beta_r = \frac{\partial W}{\partial \alpha_r}. \quad . \quad . \quad (92)$$

From eqs. (91) and (92), we arrive at the following relation for W,

$$H\left(q_r, \frac{\partial W}{\partial q_r}\right) + \frac{\partial W}{\partial t} = 0. \qquad . \qquad . \qquad (93)$$

Further, if β_r and α_r are to be generalized co-ordinates and momenta and consequently unique functions of the p_r's and q_r's, W must be so defined that ($92a$) gives one and only one value for the α_r's. But the reader will observe, on comparison, that (93) is Hamilton's partial differential equation, that eqs. ($92a$ and b) are identical with (64) and (65), and further that the condition here imposed on W is the same as in the original enunciation of the Hamilton-Jacobi theory. We are therefore led to this theory by means of an entirely new approach, and are at the same time employing a very direct and concise method.

An analogous procedure, which results in the other form of the Hamilton-Jacobi equation, may be adduced. Our goal is now to find a set of co-ordinates and momenta which are functions of the p's and q's, but not of the time; all the momenta and all the co-ordinates except one are required to be constant during the motion of the system, while the remaining co-ordinates must increase uniformly with the time. Let the co-ordinates be designated by $\beta_0, \ldots, \beta_{n-1}$ (β_0 being the varying one), and denote the corresponding momenta by $\alpha_0, \ldots, \alpha_{n-1}$. The nature of the problem implies that the transformation function (say, S) does not involve the time, and that $\bar{H}$ is accordingly equal to H. Hence, Hamilton's canonical equations become

$$\dot{\beta}_r = \frac{\partial H}{\partial \alpha_r}; \quad \dot{\alpha}_r = -\frac{\partial H}{\partial \beta_r}.$$

Since $\dot{\beta}_0 = \text{const.}$, $\dot{\beta}_r = 0$ ($r \neq 0$) and $\dot{\alpha}_r = 0$, H is a function of α_0 only. The conditions of the problem tell us nothing about the character of this function, and we shall here adopt the simplest choice, viz.,

$$H(q_r, p_r) = \alpha_0. \qquad . \qquad . \qquad . \qquad (94)$$

($90a$ and b) are now

$$(a) \quad p_r = \frac{\partial S}{\partial q_r}; \quad (b) \quad \beta_r = \frac{\partial S}{\partial \alpha_r}, \qquad . \qquad . \qquad (95)$$

and, on account of (94), S will be governed by the equation

$$H\left(q_r, \frac{\partial S}{\partial q_r}\right) = \alpha_0. \qquad . \qquad . \qquad . \qquad (96)$$

Continuing exactly as before, S must be such a function that α_0 and the α_r's ($r = 1, 2, \ldots, n - 1$) are uniquely determined by (95a). Eq. (96) shows at once that α_0 is equal to E, the total energy; $\alpha_1, \ldots, \alpha_{n-1}$ should then be uniquely determined by any $n - 1$ of equations (95a), and the nth of these equations is automatically satisfied according to (96). Again, (96) is the Hamilton-Jacobi equation, with S subjected to the usual conditions, and eqs. (95a) and (95b, $r \neq 0$) are respectively the same as (67) and (66). Our original objective has thus been attained: the second version of the Hamilton-Jacobi theory is established.

We next pass on to contact transformations in which the new co-ordinates and momenta differ from the old by infinitesimal quantities only, i.e.,

$$Q_r = q_r + \lambda g_r; \quad P_r = p_r + \lambda h_r, \quad . \quad . \quad (97)$$

where λ is an infinitesimal constant. Eq. (84a) now reads

$$\sum_r (p_r + \lambda h_r)d(q_r + \lambda g_r) - p_r dq_r = -dF, \quad . \quad (98)$$

whence
$$\sum_r \lambda(h_r dq_r + p_r dg_r) = -dF,$$

to quantities of the first order. From this equation, it is evident that $F = \lambda G$, where G is some function of the q's and p's, and we may write

$$\sum_r (h_r dq_r + p_r dg_r) = -dG,$$

or
$$\sum_r (h_r dq_r - g_r dp_r) = -d(G + \sum_r p_r g_r),$$

so that
$$\sum_r (h_r dq_r - g_r dp_r) = -dK, \quad . \quad . \quad . \quad (99)$$

if $K = G + \sum_r p_r g_r$. Whence,

$$g_r = \frac{\partial K}{\partial p_r}; \quad h_r = -\frac{\partial K}{\partial q_r}. \quad . \quad . \quad (100)$$

With this we reach the very important conclusion that *the motion of a dynamical system may be regarded as the continuous unfolding of a contact transformation*. The theorem is an immediate consequence of Hamilton's equations, as the co-ordinates and momenta after a time δt are related to the original co-ordinates and momenta by the equations

$$Q_r = q_r + \dot{q}_r \delta t; \quad P_r = p_r + \dot{p}_r \delta t$$

and
$$\dot{q}_r = \frac{\partial H}{\partial p_r}; \qquad \dot{p}_r = -\frac{\partial H}{\partial q_r}.$$

The transformation function will be given by

$$H = G + \sum_r p_r \dot{q}_r,$$

i.e.,
$$G = -(\sum_r p_r \dot{q}_r - H)$$

$$= -L.$$

Thus,
$$F = G\delta t$$

$$= -L\delta t.$$

Over a finite time interval, the transformation function will therefore adopt the form

$$-\int L dt = -W,$$

where W is the original version of Hamilton's principal function of the initial co-ordinates q_{ra}, the final co-ordinates q_{rf}, and the time t. That the q_{rf}'s and p_{rf}'s are connected with the q_{ra}'s and p_{ra}'s by means of a contact transformation, $-W$ being the transformation function, can also be inferred directly from (60a and b), which present a special case of (87). In this, we meet with an alternative proof that the motion of a dynamical system is a continuous unfolding of a contact transformation.

§8

Electrodynamics in Hamiltonian Form

Our next object of inquiry will be the investigation of the extent to which the laws of electrodynamics can be treated as variational principles and as instances of Hamilton's canonical equations. If our examination is successful, it will have important consequences in several directions. It would imply, firstly, that any theorems proved by means of the Hamilton equations can be generalized to include electrodynamics. Moreover, such an application would illustrate the wide validity both of Hamilton's equations and of variational principles.

As a preliminary example, the motion of a point charge in an electromagnetic field must be considered. The problem is of formal interest, since the forces are not derived from a potential; in fact, they involve the velocity of the particle. The Lagrangian and Hamiltonian functions will therefore be of a form somewhat different from that discussed hitherto. For convenience, we shall operate with rectangular co-ordinate systems throughout, but the results will be true in generalized co-ordinates because of their property of invariance.

The acceleration of a charged particle in an electromagnetic field is given by the Lorentz equation†

$$\dot{\mathbf{p}}' = e \left\{ \boldsymbol{\mathcal{E}} + \frac{1}{c} (\dot{\mathbf{x}} \wedge \boldsymbol{\mathcal{H}}) \right\}, \qquad . \qquad . \qquad (101)$$

where $\boldsymbol{\mathcal{E}}$ and $\boldsymbol{\mathcal{H}}$ are the electric and magnetic vectors respectively, e the charge of the particle, and $\mathbf{p}'$ its momentum without taking the field into account. $\boldsymbol{\mathcal{E}}$ and $\boldsymbol{\mathcal{H}}$ can be derived from the scalar

† We use here a slight change of notation, writing x_1, x_2, x_3 instead of x, y, z, so that the equations may be expressed more neatly. The position vector is accordingly indicated by $\mathbf{x}$.

and vector potentials ϕ and $\mathbf{A}$ with the aid of the equations

$$\left.\begin{array}{l} \boldsymbol{\mathcal{E}} = -\operatorname{grad} \phi - \dfrac{\partial \mathbf{A}}{\partial t} \\[2mm] \boldsymbol{\mathcal{H}} = \operatorname{curl} \mathbf{A} \end{array}\right\} \qquad . \qquad . \qquad (102)$$

Eq. (101) thus becomes

$$\dot{\mathbf{p}}' = e\left\{-\operatorname{grad} \phi - \frac{1}{c}\frac{\partial \mathbf{A}}{\partial t} + \frac{1}{c}(\dot{\mathbf{x}} \wedge \operatorname{curl} \mathbf{A})\right\} \qquad . \qquad . \qquad (103)$$

$$= e\left[-\operatorname{grad} \phi - \frac{1}{c}\frac{\partial \mathbf{A}}{\partial t} + \frac{1}{c}\left\{\operatorname{grad}(\dot{\mathbf{x}} \cdot \mathbf{A}) - (\dot{\mathbf{x}} \cdot \operatorname{grad})\mathbf{A}\right\}\right],$$

or, written out,

$$\dot{p}_r' = e\left\{-\frac{\partial \phi}{\partial x_r} - \frac{1}{c}\frac{\partial A_r}{\partial t} + \sum_s \frac{\dot{x}_s}{c}\frac{\partial A_s}{\partial x_r} - \sum_s \frac{\dot{x}_s}{c}\frac{\partial A_r}{\partial x_s}\right\}$$

$$= e\left\{-\frac{\partial \phi}{\partial x_r} - \frac{1}{c}\frac{dA_r}{dt} + \sum_s \frac{\dot{x}_s}{c}\frac{\partial A_s}{\partial x_r}\right\}.$$

In order to express the equations in the Lagrangian form

$$\frac{d}{dt}\left(\frac{\partial L}{\partial \dot{x}_r}\right) - \frac{\partial L}{\partial x_r} = 0,$$

we need merely adopt as our Lagrangian function

$$L = F - e\phi + e\sum_r A_r \frac{\dot{x}_r}{c}, \qquad . \qquad . \qquad (104)$$

where F is defined as in (41). The generalized momenta will now be

$$p_r = \frac{\partial L}{\partial \dot{x}_r}$$

$$= p_r' + \frac{eA_r}{c}. \qquad . \qquad . \qquad (105)$$

Finally, the Hamiltonian is still equal to the total energy, for from (47),

$$H = -L + \sum_r p_r \dot{x}_r$$

$$= \sum_r \left\{-F - eA_r \frac{\dot{x}_r}{c} + e\phi + \left(p_r' + \frac{eA_r}{c}\right)\dot{x}_r\right\}$$

$$= T + e\phi, \text{ (from (45))},$$

or, as H must be a function of the momenta and the co-ordinates,

$$H = c \sqrt{\sum_r \left(p_r - \frac{eA_r}{c} \right)^2 + m_0^2 c^2} - m_0 c^2 + e\phi. \quad . \quad (106)$$

With these interpretations of H and p_r it follows, as a result of the general theory developed in §5, that Hamilton's canonical equations hold for the case of a particle moving in an electro-magnetic field. Unlike in pure dynamics, however, H is not a homogeneous quadratic function of the momenta, even in the non-relativistic approximation.

If we are dealing with charged bodies, and not with point charges, the Lagrangian is

$$L = T + \int \rho \left(\sum_r A_r \frac{v_r}{c} - \phi \right) d\tau, \quad . \quad (107)$$

where **v** is the velocity of the charge at any point, ρ the charge density and $d\tau$ an element of volume. To obtain this equation, one need only regard the body as a system of mutually attracting particles. The v_r's and ρ are to be treated as functions of whatever generalized co-ordinates are used to define the system. According to (107), the generalized momenta will be determined by the formula

$$p_r = p_r' + \int \sum_s \rho \frac{\partial v_s}{\partial \dot{q}_r} \frac{A_s}{c} d\tau, \quad . \quad . \quad (108)$$

and the Hamiltonian may be written

$$H = H \left(q_r, p_r - \int \sum_s \rho \frac{\partial v_s}{\partial \dot{q}_r} \frac{A_s}{c} d\tau \right) + \int \rho \phi d\tau. \quad . \quad (109)$$

At this point, we may turn to the main topic of the section, that is, the actual field equations themselves.[20,21] There is a fundamental difference between this case and all instances previously considered. The reason is that until now there occurred one independent variable t and several dependent variables q_r, whereas here the x_r's and t are all independent variables, and the quantities specifying the field are the dependent variables. Suppose we have a field (not necessarily electro-magnetic) defined by the quantities $f_r (x_1, x_2, x_3, t)$. We are required to find a "Lagrangian" L which is a function of the f_r's, the $\frac{\partial f_r}{\partial x_s}$'s and the $\dot{f}_r$'s, so chosen that the differential equations (57) for the integral $\int L dx_1 dx_2 dx_3 dt$ to be stationary are the equations of the field. This variational principle can be rendered analogous to Hamilton's principle by dividing the space x_1, x_2, x_3 into a large number of small "cells" $\delta\tau$ and by supposing the field to be specified by the values of the f_r's at one arbitrarily chosen point **z** in each of these cells. We assign the symbols f_{rz} to these quantities. Let the new Lagrangian be

$$L = \sum_\tau \mathsf{L_z} \delta\tau, \qquad . \qquad . \qquad . \qquad (110)$$

in which the derivatives in $\mathsf{L_z}$ are replaced by finite differences. Hamilton's principle

$$\delta\int L\,dt = 0$$

will, as the volumes of the cells approach zero, become identical with our variational principle

$$\delta\int L\,dx_1 dx_2 dx_3 dt = 0.$$

The ordinary Lagrange equations (33) for L and eqs. (57) for L,

i.e.,
$$\frac{\partial}{\partial t}\left(\frac{\partial \mathsf{L}}{\partial \dot{f_r}}\right) + \sum_s \frac{\partial}{\partial x_s}\left(\frac{\partial \mathsf{L}}{\partial \left(\dfrac{\partial f_r}{\partial x_s}\right)}\right) - \frac{\partial \mathsf{L}}{\partial f_r} = 0, \qquad . \quad (111)$$

are therefore equivalent, as they both lead to the same variational principle. Since this equivalence might not appear so evident at first sight, it will be germane to demonstrate it directly, in order to bring the development of the theory within the scope of exact treatment. Eq. (110) may be rewritten

$$L(f_{rz}, \dot{f}_{rz}) = \sum_\tau \mathsf{L_z}\left(f_{rz}, \frac{\delta f_{rz}}{\delta x_s}, \dot{f}_{rz}\right)\delta\tau.$$

Now, in differentiating L with respect to f_{rz}, we have, as regards the terms $\dfrac{\delta f_{rz}}{\delta x_s}$, to consider both $\mathsf{L_z}$ and $\mathsf{L_{z'}}$, where $\mathsf{L_{z'}}$ is the L-function of the cell whose x_s co-ordinate is immediately above that of the cell **z**. If the difference between these x_s co-ordinates is δx_s, the term $\dfrac{\delta f_{rz}}{\delta x_s}$ will read $\dfrac{f_{rz'} - f_{rz}}{\delta x_s}$. Thus, differentiation of $\mathsf{L_{z'}}$ with respect to f_{rz} will result in a term

$$-\frac{\left\{\dfrac{\partial \mathsf{L}}{\partial\left(\dfrac{\delta f_{rz}}{\delta x_s}\right)}\right\}_{z'}}{\delta x_s}.$$

Differentiating L with respect to f_{rz} will similarly, as far as depends on $\dfrac{\delta f_{rz}}{\delta x_s}$, yield a term

$$\frac{\left\{\dfrac{\partial \mathsf{L}}{\partial \left(\dfrac{\delta f_{rz}}{\delta x_s}\right)}\right\}_z}{\delta x_s}.$$

The total contribution of terms involving $\dfrac{\delta f_{rz}}{\delta x_s}$ to the derivative of L will thus be

$$-\frac{\delta\left(\dfrac{\partial \mathsf{L}}{\partial \left(\dfrac{\delta f_{rz}}{\delta x_s}\right)}\right)}{\delta x_s},$$

the δ in the numerator meaning the increment over the distance δx_s. As the volume of the cells tends to zero, the contribution to $\dfrac{\partial L}{\partial f_{rz}}$ of the terms in $\dfrac{\partial f_r}{\partial x_s}$ will therefore be

$$-\left\{\frac{\partial}{\partial x_s}\left(\frac{\partial \mathsf{L}}{\partial \left(\dfrac{\partial f_r}{\partial x_s}\right)}\right)\right\}_z$$

and hence we may write

$$\frac{\partial L}{\partial f_{rz}} = \left\{\frac{\partial \mathsf{L}}{\partial f_r} - \sum_s \frac{\partial}{\partial x_s}\left(\frac{\partial \mathsf{L}}{\partial \left(\dfrac{\partial f_r}{\partial x_s}\right)}\right)\right\}_z \delta\tau. \qquad (112)$$

Eq. (111) then follows at once, and our proof is hereby completed. This method of differentiating with respect to f_r is sometimes known as "Hamiltonian differentiation."

Proceeding in the customary way, we define the generalized momenta corresponding to the field "co-ordinates" f_{rz} by the equations

$$p_{rz} = \frac{\partial L}{\partial \dot{f}_{rz}} \qquad \cdot \qquad \cdot \qquad \cdot \qquad (113)$$

$$= \frac{\partial \mathsf{L}_z}{\partial \dot{f}_{rz}} \delta\tau$$

$$= P_{rz}\, \delta\tau, \qquad \cdot \qquad \cdot \qquad \cdot \qquad (114)$$

where
$$P_r = \frac{\partial \mathsf{L}}{\partial \dot{f}_r}, \qquad \cdot \qquad \cdot \qquad \cdot \qquad (115)$$

and, defining the Hamiltonian as before,

$$H = \sum_{r,z} p_{rz}\dot{f}_{rz} - L \qquad . \qquad . \qquad . \qquad (116)$$

$$= \sum_{z} (\sum_{r} P_{rz}\dot{f}_{rz} - \mathsf{L}_z)\delta\tau$$

$$= \sum_{z} \mathsf{H}_z\delta\tau, \qquad . \qquad . \qquad . \qquad (117)$$

H being given by

$$\mathsf{H} = \sum_{r} P_r\dot{f}_r - \mathsf{L}. \qquad . \qquad . \qquad . \qquad (118)$$

In the limit of infinitesimal cells, eq. (117) reads

$$H = \int \mathsf{H}\, d\tau. \qquad . \qquad . \qquad . \qquad (119)$$

We are now sufficiently equipped to describe the change of the field with time by means of Hamilton's canonical equations. These are, substituting f_{rz} and p_{rz} for p_r and q_r in (48),

$$(a) \qquad \dot{f}_{rz} = \frac{\partial H}{\partial p_{rz}}$$
$$\left.\right\} \qquad . \qquad . \qquad (120)$$
$$(b) \qquad \dot{p}_{rz} = -\frac{\partial H}{\partial f_{rz}}$$

Applied to H, they transform into

$$(a) \qquad \dot{f}_r = \frac{\partial \mathsf{H}}{\partial P_r}$$
$$\left.\right\}, \qquad . \qquad (121)$$
$$(b) \qquad \dot{P}_r = -\frac{\partial \mathsf{H}}{\partial f_r} + \sum_{s} \frac{\partial}{\partial x_s}\left(\frac{\partial \mathsf{H}}{\partial \left(\frac{\partial f_r}{\partial x_s}\right)}\right)$$

when we differentiate with respect to f_{rz} in the same manner as in eq. (112). Alternatively, one can obtain (121) by direct transformation of (111), pursuing the same method as in pure dynamics. Proceeding accordingly, we get from eq. (118),

$$\mathsf{H} = \sum_{r} P_r\dot{f}_r - \mathsf{L},$$

so that
$$\frac{\partial H}{\partial P_r} = \dot{f}_r + \sum_s P_s \frac{\partial \dot{f}_s}{\partial P_r} - \sum_s \frac{\partial L}{\partial \dot{f}_s} \frac{\partial \dot{f}_s}{\partial P_r}$$

$$= \dot{f}_r + \sum_s P_s \frac{\partial \dot{f}_s}{\partial P_r} - \sum_s P_s \frac{\partial \dot{f}_s}{\partial P_r}$$

$$= \dot{f}_r;$$

$$\frac{\partial H}{\partial f_r} - \sum_s \frac{\partial}{\partial x_s}\left(\frac{\partial H}{\partial\left(\frac{\partial f_r}{\partial x_s}\right)}\right)$$

$$= \sum_t P_t \left(\frac{\partial \dot{f}_t}{\partial f_r}\right)_P - \sum_{s,t} \frac{\partial}{\partial x_s}\left\{ P_t \left(\frac{\partial \dot{f}_t}{\partial\left(\frac{\partial f_r}{\partial x_s}\right)}\right)_P \right\}$$

$$- \left(\frac{\partial L}{\partial f_r}\right)_P + \sum_s \frac{\partial}{\partial x_s}\left(\frac{\partial L}{\partial\left(\frac{\partial f_r}{\partial x_s}\right)}\right)_P$$

$$= \sum_t P_t \left(\frac{\partial \dot{f}_t}{\partial f_r}\right)_P - \sum_{s,t} \frac{\partial}{\partial x_s}\left\{ P_t \left(\frac{\partial \dot{f}_t}{\partial\left(\frac{\partial f_r}{\partial x_s}\right)}\right)_P \right\}$$

$$- \left(\frac{\partial L}{\partial f_r}\right)_f - \sum_t \left(\frac{\partial L}{\partial \dot{f}_t}\right)_f \left(\frac{\partial \dot{f}_t}{\partial f_r}\right)_P$$

$$+ \sum_s \frac{\partial}{\partial x_s}\left(\frac{\partial L}{\partial\left(\frac{\partial f_r}{\partial x_s}\right)}\right)_f + \sum_{s,t} \frac{\partial}{\partial x_s}\left\{ \left(\frac{\partial L}{\partial \dot{f}_t}\right)_f \left(\frac{\partial \dot{f}_t}{\partial\left(\frac{\partial f_r}{\partial x_s}\right)}\right)_P \right\}$$

$$= - \left(\frac{\partial L}{\partial f_r}\right)_f + \sum_s \frac{\partial}{\partial x_s}\left(\frac{\partial L}{\partial\left(\frac{\partial f_r}{\partial x_s}\right)}\right), \text{ as } P_t = \frac{\partial L}{\partial \dot{f}_t};$$

$$= - \frac{\partial}{\partial t}\left(\frac{\partial L}{\partial \dot{f}_r}\right), \text{ from (111)},$$

$$= - \frac{\partial P_r}{\partial t}.$$

As in the theory of contact transformations in §7, we observe that (121) is the condition for $\int \left\{ \sum\limits_r P_r \dot{f}_r - \mathsf{H}\left(f_r, \dfrac{\partial f_r}{\partial x_s}, P_r\right) \right\} d\tau \, dt$ to have a stationary value, the P's and f's being regarded as entirely independent of one another. Now, it is always possible to describe the field and its time rate of change by means of new variables $P_s{}'$, $f_s{}'$ (e.g., quantities related to the Fourier resolution of the field), which may be functions of a set of discrete or continuous variables k_t (the x_1-, x_2-, x_3-components of the wave-length, in the case of the Fourier resolution). The Hamilton equations will then hold in the new system, if

$$\int \sum_r P_r \dot{f}_r d\tau = \int \left(\sum_s P_s{}' \dot{f}_s{}' - \frac{dF}{dt} \right) dk_1 dk_2 \ldots, \qquad (122)$$

where F signifies any function of the f''s, the $\dfrac{\partial f'}{\partial k_t}$'s, and the P''s. In the case of the k_t's being discrete variables, eq. (122) must be substituted by

$$\int \sum_r P_r \dot{f}_r d\tau = \sum_{s,k_t} \{(P_s)_{k_t} (\dot{f}_s)_{k_t}\} - \frac{dF}{dt}. \qquad (123)$$

We shall therefore call the transformation to these new variables a contact transformation. The conditions (122) or (123) for a contact transformation also follow, on replacing, as before, the continuous field variables by a large number of discrete variables and applying the theory worked out in §7.

To test whether this general theory is adjustable to any particular case, one must first investigate whether there exists a function L for which (111) are the equations of the field. Our next step is thus to find such a function for the electromagnetic field *in vacuo*. The f_r's are then the vector potential **A** and the scalar potential ϕ, and the field equations take the form

$$(a) \qquad \nabla^2 \mathbf{A} - \frac{1}{c^2} \frac{\partial^2 \mathbf{A}}{\partial t^2} = 0 \left.\begin{array}{c} \\ \\ \end{array}\right\}$$
$$\qquad\qquad\qquad\qquad\qquad\qquad\qquad . \quad . \quad . \quad (124)$$
$$(b) \qquad \nabla^2 \phi - \frac{1}{c^2} \frac{\partial^2 \phi}{\partial t^2} = 0 \left.\begin{array}{c} \\ \\ \end{array}\right.$$

The potentials have to be subjected to the further restriction that

$$\operatorname{div} \mathbf{A} + \frac{1}{c} \frac{\partial \phi}{\partial t} = 0, \qquad . \qquad . \qquad (125)$$

which is best interpreted as a prerequisite to be added over and above the equations of motion, as this equation is satisfied for all time, once it and its first derivative with respect to t hold initially. A Lagrangian which gives rise to the required equations (124) is easily obtained, viz.,

$$\mathsf{L} = \frac{1}{8\pi} \left(\mathcal{E}^2 - \mathcal{H}^2 \right). \qquad \cdot \qquad \cdot \qquad (126)$$

The factor $1/8\pi$ is, of course, completely arbitrary and is only inserted so as to facilitate the later application to the theory of the electromagnetic field in interaction with charged bodies. $\mathcal{H}$ and $\mathcal{E}$ in (126) are to be expressed in terms of the potentials, i.e.,

$$\mathsf{L} = \frac{1}{8\pi} \left\{ \left(-\mathrm{grad}\, \phi - \frac{1}{c} \frac{\partial \mathbf{A}}{\partial t} \right)^2 - (\mathrm{curl}\, \mathbf{A})^2 \right\} \qquad \cdot \qquad \cdot \qquad (127)$$

$$= \frac{1}{8\pi} \left\{ \left(-\frac{\partial \phi}{\partial x_1} - \frac{1}{c} \frac{\partial A_1}{\partial t} \right)^2 + \left(-\frac{\partial \phi}{\partial x_2} - \frac{1}{c} \frac{\partial A_2}{\partial t} \right)^2 \right.$$

$$+ \left(-\frac{\partial \phi}{\partial x_3} - \frac{1}{c} \frac{\partial A_3}{\partial t} \right)^2 - \left(\frac{\partial A_3}{\partial x_2} - \frac{\partial A_2}{\partial x_3} \right)^2$$

$$\left. - \left(\frac{\partial A_1}{\partial x_3} - \frac{\partial A_3}{\partial x_1} \right)^2 - \left(\frac{\partial A_2}{\partial x_1} - \frac{\partial A_1}{\partial x_2} \right)^2 \right\}. \qquad \cdot \qquad \cdot \qquad (128)$$

Then

$$\frac{\partial}{\partial t} \left(\frac{\partial \mathsf{L}}{\partial \left(\frac{\partial A_1}{\partial t} \right)} \right) + \sum_r \left\{ \frac{\partial}{\partial x_r} \left(\frac{\partial \mathsf{L}}{\partial \left(\frac{\partial A_1}{\partial x_r} \right)} \right) \right\} - \frac{\partial \mathsf{L}}{\partial A_1}$$

$$= \frac{1}{4\pi} \left\{ \frac{1}{c} \frac{\partial}{\partial t} \left(\frac{\partial \phi}{\partial x_1} + \frac{1}{c} \frac{\partial A_1}{\partial t} \right) + \frac{\partial}{\partial x_2} \left(\frac{\partial A_2}{\partial x_1} - \frac{\partial A_1}{\partial x_2} \right) \right.$$

$$\left. - \frac{\partial}{\partial x_3} \left(\frac{\partial A_1}{\partial x_3} - \frac{\partial A_3}{\partial x_1} \right) \right\}$$

$$= \frac{1}{4\pi} \left\{ -\nabla^2 A_1 + \frac{1}{c^2} \frac{\partial^2 A_1}{\partial t^2} + \frac{\partial}{\partial x_1} \left(\mathrm{div}\, \mathbf{A} + \frac{1}{c} \frac{\partial \phi}{\partial t} \right) \right\}$$

$$= -\frac{1}{4\pi} \left(\nabla^2 A_1 - \frac{1}{c^2} \frac{\partial^2 A_1}{\partial t^2} \right), \qquad \cdot \qquad \cdot \qquad \cdot \qquad \cdot \qquad (129)$$

using (125).†

† It is of interest to notice that the Lagrangian (126) can be used even if we do not restrict the potentials by the supplementary condition, since the equations we obtain before applying (125) are those which the unrestricted potentials would satisfy. Such unrestricted potentials are, however, not of great importance.

Similar equations can be derived for A_2 and A_3. Finally,

$$\frac{\partial}{\partial t}\left(\frac{\partial \mathsf{L}}{\partial\left(\frac{\partial \phi}{\partial t}\right)}\right) + \sum_r \left\{\frac{\partial}{\partial x_r}\left(\frac{\partial \mathsf{L}}{\partial\left(\frac{\partial \phi}{\partial x_r}\right)}\right)\right\} - \frac{\partial \mathsf{L}}{\partial \phi}$$

$$= \frac{1}{4\pi}\left\{\frac{\partial}{\partial x_1}\left(\frac{\partial \phi}{\partial x_1} + \frac{1}{c}\frac{\partial A_1}{\partial t}\right) + \frac{\partial}{\partial x_2}\left(\frac{\partial \phi}{\partial x_2} + \frac{1}{c}\frac{\partial A_2}{\partial t}\right)\right.$$

$$\left. + \frac{\partial}{\partial x_3}\left(\frac{\partial \phi}{\partial x_3} + \frac{1}{c}\frac{\partial A_3}{\partial t}\right)\right\}$$

$$= \frac{1}{4\pi}\left\{\nabla^2\phi - \frac{1}{c^2}\frac{\partial^2\phi}{\partial t^2} + \frac{1}{c}\frac{\partial}{\partial t}\left(\operatorname{div}\mathbf{A} + \frac{1}{c}\frac{\partial \phi}{\partial t}\right)\right\}$$

$$= \frac{1}{4\pi}\left(\nabla^2\phi - \frac{1}{c^2}\frac{\partial^2\phi}{\partial t^2}\right), \qquad \cdot \qquad \cdot \qquad \cdot \qquad \cdot \qquad (130)$$

again from (125). The Lagrangian (126) does therefore lead to the correct equations of the field. Electrodynamics of a vacuum can consequently be treated with the aid of variational principles and Hamiltonian mechanics, at any rate in the sense discussed in this section.

In the following pages, it will be preferable to operate with a Lagrangian which furnishes eqs. (124) immediately, without invoking (125): it is desirable to conceive of this last equation as an additional requirement which the potentials must obey and of which eqs. (124) are independent. Such a Lagrangian is given by

$$\mathsf{L} = \frac{1}{8\pi}\left\{\left(-\operatorname{grad}\phi - \frac{1}{c}\frac{\partial \mathbf{A}}{\partial t}\right)^2\right.$$

$$\left. - (\operatorname{curl}\mathbf{A})^2 - \left(\operatorname{div}\mathbf{A} + \frac{1}{c}\frac{\partial \phi}{\partial t}\right)^2\right\}, \qquad \cdot \qquad (131)$$

as the reader may verify without difficulty.

We shall not advance any further this analysis of the electromagnetic field in a vacuum, but pass on to a field in interaction with charged bodies, of which the vacuum field is a special case. The potentials then comply with the equations

$$(a) \qquad \nabla^2\mathbf{A} - \frac{1}{c^2}\frac{\partial^2\mathbf{A}}{\partial t^2} = -4\pi\rho\,\frac{\mathbf{v}}{c}$$

$$(b) \qquad \nabla^2\phi - \frac{1}{c^2}\frac{\partial^2\phi}{\partial t^2} = -4\pi\rho \qquad\qquad (132)$$

The Lagrangian will result in these equations if we add to (131) the term $\rho\left(\mathbf{A}\cdot\dfrac{\mathbf{v}}{c} - \phi\right)$, where ρ and $\mathbf{v}$ are the density and velocity of the charge respectively, so that

$$\mathsf{L} = \frac{1}{8\pi}\left\{\left(-\operatorname{grad}\phi - \frac{1}{c}\frac{\partial\mathbf{A}}{\partial t}\right)^2 - (\operatorname{curl}\mathbf{A})^2\right.$$

$$\left. - \left(\operatorname{div}\mathbf{A} + \frac{1}{c}\frac{\partial\phi}{\partial t}\right)^2\right\} + \rho\left(\mathbf{A}\cdot\frac{\mathbf{v}}{c} - \phi\right) \quad (133)$$

There would now be another term $-\rho\dfrac{v_1}{c}$ on the right side of (129), and a term ρ on the right side of (130). Hence we are able to deduce eqs. (132). The function L for the field will, according to (110), be obtained by integrating (122) over the whole space. But on comparison we notice that the inter-action term in this function (i.e., the term depending on the field variables *and* the co-ordinates of the dynamical system) is exactly the same as the interaction term in (107). This leads to the remarkable result that, if we assume a Lagrangian for the combined dynamical system and field, given by

$$L = T + \int\left[\frac{1}{8\pi}\left\{\left(-\operatorname{grad}\phi - \frac{1}{c}\frac{\partial\mathbf{A}}{\partial t}\right)^2 - (\operatorname{curl}\mathbf{A})^2\right.\right.$$

$$\left.\left. - \left(\operatorname{div}\mathbf{A} + \frac{1}{c}\frac{\partial\phi}{\partial t}\right)^2\right\} + \rho\left(\mathbf{A}\cdot\frac{\mathbf{v}}{c} - \phi\right)\right] d\tau, \quad (134)$$

an application of Hamilton's principle will embrace both the equations of motion of the system and the equations of the electromagnetic field. On replacing the continuous field variables by a set of discrete variables in conformity with our usual method, the Lagrange equations with the above expression for L (the integral being substituted by a summation) are equivalent to our required dynamical and electromagnetic equations.

In accordance with (115), the quantities P conjugate to the field variables will be

$$P_{A_r} = \frac{\partial \mathsf{L}}{\partial \dot{A}_r}$$

$$= \frac{1}{4\pi c} \left(\frac{\partial \phi}{\partial x_r} + \frac{1}{c} \frac{\partial A_r}{\partial t} \right)$$

$$= -\frac{\mathcal{E}_r}{4\pi c} . \qquad . \qquad . \qquad . \qquad . \qquad (135)$$

In the same way, $P_\phi = \dfrac{\partial \mathsf{L}}{\partial \dot{\phi}}$

$$= -\frac{1}{4\pi c} \left(\operatorname{div} \mathbf{A} + \frac{1}{c} \frac{\partial \phi}{\partial t} \right). \qquad . \qquad . \qquad (136)$$

We are now in a position to compute the Hamiltonian H. The contribution due to the first and the last terms of (134) and to the sum $\sum_r p_r \dot{q}_r$ for the dynamical system is determined by (109). The remaining contribution will be that of the terms in the field quantities alone, i.e.,

$$H = \int \mathsf{H} d\tau,$$

where

$$\mathsf{H} = \sum P_{A_r} \dot{A}_r + P_\phi \dot{\phi} - \frac{1}{8\pi} \left\{ \left(-\operatorname{grad} \phi - \frac{1}{c} \frac{\partial \mathbf{A}}{\partial t} \right)^2 - (\operatorname{curl} \mathbf{A})^2 \right.$$

$$\left. - \left(\operatorname{div} \mathbf{A} + \frac{1}{c} \frac{\partial \phi}{\partial t} \right)^2 \right\}$$

$$= \frac{1}{8\pi} \left\{ -\frac{2}{c} \mathcal{E} \frac{\partial \mathbf{A}}{\partial t} - \frac{2}{c} \left(\operatorname{div} \mathbf{A} + \frac{1}{c} \frac{\partial \phi}{\partial t} \right) \frac{\partial \phi}{\partial t} - \mathcal{E}^2 + \mathcal{H}^2 \right.$$

$$\left. + \left(\operatorname{div} \mathbf{A} + \frac{1}{c} \frac{\partial \phi}{\partial t} \right)^2 \right\}$$

$$= \frac{1}{8\pi} \left\{ 2\mathcal{E}^2 + 2\mathcal{E} \cdot \operatorname{grad} \phi \right.$$

$$\left. + \left(\operatorname{div} \mathbf{A} + \frac{1}{c} \frac{\partial \phi}{\partial t} \right) \left(\operatorname{div} \mathbf{A} - \frac{1}{c} \frac{\partial \phi}{\partial t} \right) - \mathcal{E}^2 + \mathcal{H}^2 \right\}$$

$$= \frac{1}{8\pi} \left[\mathcal{E}^2 + \mathcal{H}^2 + 2\mathcal{E} \cdot \operatorname{grad} \phi + \left\{ (\operatorname{div} \mathbf{A})^2 - \frac{1}{c^2} \left(\frac{\partial \phi}{\partial t} \right)^2 \right\} \right].$$

$$(137)$$

Hence $H = H_s \left(q_r, p_r - \int \sum_s \rho \, \frac{\partial v_s}{\partial \dot{q}_r} \frac{A_s}{c} \, d\tau \right)$

$$+ \int \left[\rho \phi + \frac{1}{8\pi} \left\{ \mathcal{E}^2 + \mathcal{H}^2 + 2\mathcal{E} \cdot \operatorname{grad} \phi + (\operatorname{div} \mathbf{A})^2 \right. \right.$$

$$\left. \left. - \frac{1}{c^2} \left(\frac{\partial \phi}{\partial t} \right)^2 \right\} \right] d\tau. \qquad \cdot \qquad \cdot \qquad \cdot \qquad (138)$$

H_s is the Hamiltonian function of the dynamical system. Setting H as a function of the field quantities, their space derivatives and the conjugate quantities P, eq. (138) will read

$$H = H_s \left(q_r, p_r - \int \sum_s \rho \, \frac{\partial v_s}{\partial \dot{q}_r} \frac{A_s}{c} \, d\tau \right)$$

$$+ \int \left[\rho \phi + \frac{1}{8\pi} \left\{ 16\pi^2 \, c^2 \sum_r P_{A_r}^{\ 2} + (\operatorname{curl} \mathbf{A})^2 - 8\pi c \sum_r P_{A_r} \frac{\partial \phi}{\partial x_r} \right. \right.$$

$$\left. \left. - 4\pi c P_\phi \left(2 \operatorname{div} \mathbf{A} + 4\pi c P_\phi \right) \right\} \right] d\tau. \qquad \cdot \qquad \cdot \qquad \cdot \qquad \cdot \qquad (139)$$

The terms which involve the field variables are not merely the integral of a function H over the volume, but, on expanding H_s, some of the integrals will be squared. It is therefore necessary to investigate the manner in which the electromagnetic equations can be deduced from such a Hamiltonian. When the field variables are taken as discrete, the ordinary Hamilton canonical equations can be used for the dynamical as well as for the electromagnetic variables, with the p's given by (114); again, the integral sign in (139) must be replaced by a summation sign. When we pass to the limit, it is clear that, in the case of continuous field variables, the differentiation with respect to A_r of the first term of (139) has to be carried out as follows: the integral sign under which the A_r stands is omitted, but any coefficient of this integral sign is written down *in toto*, even though this coefficient itself may include an integral among its terms. As for the rest of (139), we simply denote the function under the integral sign by H and differentiate according to (121) as before. On working out the Hamilton equations in this way, they will lead us back to (135), (136) and (131), as may easily be verified.

It remains to prove that, after applying the supplementary conditions (125), the Hamiltonian (138) is equal to the total energy. The first term in this Hamiltonian is the kinetic energy of the dynamical system. The sum of the last two terms under the integral sign is zero, owing to the supplementary conditions. The first and fourth terms under the integral sign cancel out, since

$$\int (\rho \phi + \frac{1}{4\pi} \mathcal{E} \cdot \operatorname{grad} \phi) \, d\tau = \int \left(\frac{1}{4\pi} \phi \operatorname{div} \mathcal{E} + \frac{1}{4\pi} \mathcal{E} \cdot \operatorname{grad} \phi \right) d\tau$$

$$= \int \left\{ \frac{1}{4\pi} \operatorname{div} (\phi \mathcal{E}) \right\} d\tau$$

$$= 0,$$

if the potentials vanish sufficiently rapidly at infinity. We are left with

$$\int \frac{1}{8\pi} (\mathcal{E}^2 + \mathcal{H}^2) \, d\tau,$$

which is the energy of the electromagnetic field.

A set of variables which is frequently used, especially in the quantum theory of radiation, to characterize an electromagnetic field is represented by the Fourier components of the scalar and vector potentials. With these variables, the Hamiltonian function and the Hamilton canonical equations assume a neater form, and certain significant conclusions can be reached.

Suppose

$$A_r = \int_{-\infty}^{\infty} (A_{k_1, k_2, k_3})_r \, e^{i(k_1 x_1 + k_2 x_2 + k_3 x_3)} \, dk_1 dk_2 dk_3,$$

or, more concisely,

$$\left. \begin{aligned} A_r &= \int_{-\infty}^{\infty} (A_{\mathbf{k}})_r \, e^{i\mathbf{k} \cdot \mathbf{x}} d^3\mathbf{k} \\ \text{and} \quad \phi &= \int_{-\infty}^{\infty} \phi_{\mathbf{k}} \, e^{i\mathbf{k} \cdot \mathbf{x}} d^3\mathbf{k} \end{aligned} \right\} \qquad \cdot \quad \cdot \quad \cdot \quad \cdot \quad (140)$$

Then the quantities $(A_{\mathbf{k}})_r$ and $\phi_{\mathbf{k}}$ are our new variables. The electric and magnetic fields can now be expressed in a Fourier series as

$$\boldsymbol{\mathcal{E}} = -\text{grad } \phi - \frac{1}{c}\frac{\partial \mathbf{A}}{\partial t}$$

$$= \int \left(-i\mathbf{k}\phi_{\mathbf{k}} - \frac{1}{c}\dot{\mathbf{A}}_{\mathbf{k}}\right)e^{i\mathbf{k}.\mathbf{x}}d^3\mathbf{k}; \qquad . \qquad . \qquad . \quad (141)$$

$$\boldsymbol{\mathcal{H}} = \text{curl } \mathbf{A}$$

$$= \int i\mathbf{k}\wedge\mathbf{A}\,e^{i\mathbf{k}.\mathbf{x}}d^3\mathbf{k}. . \qquad . \qquad . \qquad . \qquad . \qquad . \quad (142)$$

In order to determine generalized momenta, the expression $\int(\sum_r P_{A_r}\dot{A}_r + P_\phi\dot\phi)d\tau$ must be calculated in terms of the Fourier variables. Proceeding accordingly,

$$\int \sum_r (P_{A_r}\dot{A}_r + P_\phi\dot\phi)d\tau = -\frac{1}{4\pi c}\int \left\{\boldsymbol{\mathcal{E}}.\dot{\mathbf{A}} + \left(\text{div }\mathbf{A} + \frac{1}{c}\frac{\partial\phi}{\partial t}\right)\dot\phi\right\}d\tau$$

$$= -\frac{1}{4\pi c}\int\left[\left\{\int\left(-i\mathbf{k}\phi_{\mathbf{k}} - \frac{1}{c}\dot{\mathbf{A}}_{\mathbf{k}}\right)e^{i\mathbf{k}.\mathbf{x}}d^3\mathbf{k}\right\}\cdot\left\{\int\dot{\mathbf{A}}_{\mathbf{k}}e^{i\mathbf{k}.\mathbf{x}}d^3\mathbf{k}\right\}\right.$$

$$\left. + \left\{\int\left(i\mathbf{k}.\mathbf{A}_{\mathbf{k}} + \frac{1}{c}\dot\phi_{\mathbf{k}}\right)e^{i\mathbf{k}.\mathbf{x}}d^3\mathbf{k}\right\}\cdot\left\{\int\dot\phi_{\mathbf{k}}e^{i\mathbf{k}.\mathbf{x}}d^3\mathbf{k}\right\}\right]d\tau$$

$$= -\frac{1}{4\pi c}\cdot 8\pi^3\int\left(-i\dot\phi_{\mathbf{k}}\mathbf{k}.\dot{\mathbf{A}}_{(-\mathbf{k})} - \frac{1}{c}\dot{\mathbf{A}}_{\mathbf{k}}.\dot{\mathbf{A}}_{(-\mathbf{k})} + i\mathbf{k}.\mathbf{A}_{\mathbf{k}}\dot\phi_{(-\mathbf{k})}\right.$$

$$\left. + \frac{1}{c}\dot\phi_{\mathbf{k}}\dot\phi_{(-\mathbf{k})}\right)d^3\mathbf{k}$$

$$= \frac{2\pi^2}{c^2}\int\left\{\dot{\mathbf{A}}_{\mathbf{k}}.\dot{\mathbf{A}}_{(-\mathbf{k})} - \dot\phi_{\mathbf{k}}\dot\phi_{(-\mathbf{k})} + ic\frac{d}{dt}\left(\phi_{\mathbf{k}}\mathbf{k}.\mathbf{A}_{(-\mathbf{k})}\right)\right\}d^3\mathbf{k}. \quad (143)$$

It therefore follows from (122) that appropriate quantities P conjugate to the Fourier components are

$$P_{(A_{\mathbf{k}})_r} = \frac{2\pi^2}{c^2}\left(\dot{A}_{(-\mathbf{k})}\right)_r; \quad P_{(\phi_{\mathbf{k}})} = -\frac{2\pi^2}{c^2}\dot\phi_{(-\mathbf{k})}. \qquad . \quad (144)$$

The next step is the derivation of the Hamiltonian as a

function of the Fourier variables. The first term of eq. (138) and the first term under the integral sign remain the same as before, provided we regard A_r and ϕ as functions of the Fourier variables in accordance with (140). The other terms in (138) take the form

$$H - H_s = \frac{1}{8\pi}\int\left\{\boldsymbol{\mathcal{E}}^2 + \boldsymbol{\mathcal{H}}^2 + 2\,\boldsymbol{\mathcal{E}}\cdot\text{grad}\,\phi + (\text{div }\mathbf{A})^2 - \frac{1}{c^2}\left(\frac{\partial\phi}{\partial t}\right)^2\right\}d\tau$$

$$= \frac{1}{8\pi}\int\Bigg[\left\{\int\left(-i\mathbf{k}\phi_\mathbf{k} - \frac{1}{c}\dot{\mathbf{A}}_\mathbf{k}\right)e^{i\mathbf{k}\cdot\mathbf{x}}d^3\mathbf{k}\right\}^2 + \left\{\int i\mathbf{k}\wedge\mathbf{A}_\mathbf{k}e^{i\mathbf{k}\cdot\mathbf{x}}d^3\mathbf{k}\right\}^2$$

$$+ 2\left\{\int\left(-i\mathbf{k}\phi_\mathbf{k} - \frac{1}{c}\dot{\mathbf{A}}_\mathbf{k}\right)e^{i\mathbf{k}\cdot\mathbf{x}}d^3\mathbf{k}\right\}\left\{\int i\mathbf{k}\phi_\mathbf{k}e^{i\mathbf{k}\cdot\mathbf{x}}d^3\mathbf{k}\right\}$$

$$+ \left\{\int i\mathbf{k}\cdot\mathbf{A}_\mathbf{k}e^{i\mathbf{k}\cdot\mathbf{x}}d^3\mathbf{k}\right\}^2 - \frac{1}{c^2}\left\{\int\dot{\phi}_\mathbf{k}e^{i\mathbf{k}\cdot\mathbf{x}}d^3\mathbf{k}\right\}^2\Bigg]\,d\tau$$

$$= \frac{1}{8\pi}\cdot 8\pi^3\int\left\{\mathbf{k}^2\phi_\mathbf{k}\phi_{(-\mathbf{k})} + \frac{2i}{c}\,\phi_\mathbf{k}\mathbf{k}\cdot\dot{\mathbf{A}}_{(-\mathbf{k})} + \frac{1}{c^2}\dot{\mathbf{A}}_\mathbf{k}\cdot\dot{\mathbf{A}}_{(-\mathbf{k})}\right.$$

$$+ \mathbf{k}^2\mathbf{A}_\mathbf{k}\cdot\mathbf{A}_{(-\mathbf{k})} - \left(\mathbf{k}\cdot\mathbf{A}_\mathbf{k}\right)\left(\mathbf{k}\cdot\mathbf{A}_{(-\mathbf{k})}\right) - 2\mathbf{k}^2\phi_\mathbf{k}\phi_{(-\mathbf{k})}$$

$$\left.- \frac{2i}{c}\,\phi_\mathbf{k}\mathbf{k}\cdot\dot{\mathbf{A}}_{(-\mathbf{k})} + \left(\mathbf{k}\cdot\mathbf{A}_\mathbf{k}\right)\left(\mathbf{k}\cdot\mathbf{A}_{(-\mathbf{k})}\right) - \frac{1}{c^2}\dot{\phi}_\mathbf{k}\dot{\phi}_{(-\mathbf{k})}\right\}d^3\mathbf{k}$$

$$= \pi^2\int\left(\mathbf{k}^2\mathbf{A}_\mathbf{k}\cdot\mathbf{A}_{(-\mathbf{k})} + \frac{1}{c^2}\dot{\mathbf{A}}_\mathbf{k}\cdot\dot{\mathbf{A}}_{(-\mathbf{k})} - \mathbf{k}^2\phi_\mathbf{k}\phi_{(-\mathbf{k})}\right.$$

$$\left.- \frac{1}{c^2}\dot{\phi}_\mathbf{k}\dot{\phi}_{(-\mathbf{k})}\right)d^3\mathbf{k},$$

or finally,

$$H = H_s\left(q_r,\, p_r - \int\sum_s\rho\,\frac{\partial v_s}{\partial\dot{q}_r}\frac{A_s}{c}\,d\tau\right) + \int\rho\phi\,d\tau$$

$$+ \pi^2\int\left\{\sum_r\mathbf{k}^2(A_\mathbf{k})_r\cdot(A_{(-\mathbf{k})})_r + \frac{1}{c^2}\sum_r(\dot{A}_\mathbf{k})_r(\dot{A}_{(-\mathbf{k})})_r\right.$$

$$\left.- \mathbf{k}^2\phi_\mathbf{k}\phi_{(-\mathbf{k})} - \frac{1}{c^2}\dot{\phi}_\mathbf{k}\dot{\phi}_{(-\mathbf{k})}\right\}d^3\mathbf{k}, \qquad . \qquad . \qquad (145)$$

which gives, when expressed in terms of the field variables and their conjugates,

$$
\begin{aligned}
H = H_s &\left(q_r, \, p_r - \int \sum_s \rho \, \frac{\partial v_s}{\partial \dot{q}_r} \frac{A_s}{c} \, d\tau \right) + \int \rho \phi \, d\tau \\
&+ \int \left\{ \pi^2 \mathbf{k}^2 \sum_r (A_{\mathbf{k}})_r (A_{(-\mathbf{k})})_r \right. \\
&+ \frac{c^2}{4\pi^2} \sum_r P_{(A_{\mathbf{k}})_r} P_{(A_{(-\mathbf{k})})_r} - \pi^2 \mathbf{k}^2 \phi_{\mathbf{k}} \phi_{(-\mathbf{k})} \\
&\left. - \frac{c^2}{4\pi^2} P_{\phi_{\mathbf{k}}} P_{\phi_{(-\mathbf{k})}} \right\} d^3\mathbf{k}. \qquad . \qquad . \qquad (146)
\end{aligned}
$$

It may be appropriate to verify that the Hamilton equations with the Hamiltonian function (146) do lead to eq. (132). With the help of the equation

$$
\rho \mathbf{V} = \frac{1}{8\pi^3} \int \left\{ \int \rho \mathbf{v} e^{i\mathbf{k} \cdot \mathbf{x}} \, d\tau \right\} e^{i(-\mathbf{k}) \cdot \mathbf{x}} d^3\mathbf{k},
$$

(132) can be rewritten

$$
\left. \begin{aligned}
(a) \qquad & \mathbf{k}^2 \mathbf{A}_{(-\mathbf{k})} + \frac{1}{c^2} \frac{\partial^2 \mathbf{A}_{(-\mathbf{k})}}{\partial t^2} = \frac{1}{2\pi^2 c} \int \rho \mathbf{v} e^{i\mathbf{k} \cdot \mathbf{x}} \, d\tau \\
(b) \qquad & \mathbf{k}^2 \phi_{(-\mathbf{k})} + \frac{1}{c^2} \frac{\partial^2 \phi_{(-\mathbf{k})}}{\partial t^2} = \frac{1}{2\pi^2} \int \rho e^{i\mathbf{k} \cdot \mathbf{x}} \, d\tau
\end{aligned} \right\} . \qquad . \qquad (147)
$$

Eqs. (121a), with the Hamiltonian now under consideration, are immediately equivalent to (144). On applying the second set of the Hamilton equations according to the method discussed on page 84, we get

$$
-\dot{P}_{(A_{\mathbf{k}})_r} = -\sum_t \int \rho \, \frac{\partial v_r}{\partial \dot{q}_t} \cdot \frac{1}{c} \, e^{i\mathbf{k} \cdot \mathbf{x}} \, d\tau \, \frac{\partial H_s}{\partial p_t} \\
+ 2\pi^2 \mathbf{k}^2 (A_{(-\mathbf{k})})_r. \qquad . \qquad . \qquad . \qquad (148)
$$

The first term on the right is obtained from H_s by omitting the sign which represents the integral with respect to $\mathbf{k}$ in the usual way, while the second term results from the field Hamiltonian, in which each term $\mathbf{A}_{\mathbf{k}}$ occurs twice. Thus, after substitution of $\dot{q}_t$ for $\partial H_s / \partial p_t$ and simplification, eq. (148) adopts the form

$$
- \frac{2\pi^2}{c^2} \frac{\partial^2 (A_{(-\mathbf{k})})_r}{\partial t^2} = - \frac{1}{c} \int \rho \mathbf{v} e^{i\mathbf{k} \cdot \mathbf{x}} \, d\tau + 2\pi^2 \mathbf{k}^2 (A_{(-\mathbf{k})})_r,
$$

which is the same as (147a). Similarly,

$$-\dot{P}_{\phi_{\mathbf{k}}} = \int \rho e^{i\mathbf{k}.\mathbf{x}}\, d\tau - 2\pi^2 \mathbf{k}^2 \phi_{(-\mathbf{k})},$$

i.e., $\quad \dfrac{2\pi^2}{c^2} \dfrac{\partial^2 \phi_{(-\mathbf{k})}}{\partial t^2} = \int \rho e^{i\mathbf{k}.\mathbf{x}}\, d\tau - 2\pi^2 \mathbf{k}^2 \phi_{(-\mathbf{k})},$

which is equivalent to (147b).

In order to state the restrictions which are to be imposed upon the variables so that they comply with the supplementary condition (125), the vector potential must be expanded as "transverse" and "longitudinal" waves. In other words, corresponding to each direction $\mathbf{k}$, we choose two directions λ_1 and λ_2 perpendicular to each other and to $\mathbf{k}$. We now change to the variables $(A_{\mathbf{k}})_{\lambda_1}$, $(A_{\mathbf{k}})_{\lambda_2}$ and $(A_{\mathbf{k}})_{\sigma}$, which are the coefficients of $e^{i\mathbf{k}.\mathbf{x}}$ in the Fourier analysis of the λ_1-, λ_2- and $\mathbf{k}$-components of the vector potential respectively; the λ-waves are transverse, the σ-waves longitudinal. A brief inspection will show that the transverse waves are solenoidal vectors, whereas the longitudinal waves are irrotational vectors. As

$$\operatorname{div} (A_{\mathbf{k}})_\sigma e^{i\mathbf{k}.\mathbf{x}} = i\,|\,\mathbf{k}\,|\,(A_{\mathbf{k}})_\sigma e^{i\mathbf{k}.\mathbf{x}},$$

the supplementary condition (125) may be expressed

$$i\,|\,\mathbf{k}\,|\,(A_{\mathbf{k}})_\sigma + \frac{1}{c}\frac{\partial \phi_{\mathbf{k}}}{\partial t} = 0. \qquad . \qquad . \qquad (149)$$

It has already been stated that eq. (125) and its first derivative with respect to time must both be satisfied initially; under these conditions, (125) will continue to hold. From (149), we therefore find as our required restrictions

$$\left.\begin{aligned} i\,|\,\mathbf{k}\,|\,(A_{\mathbf{k}})_\sigma + \frac{\partial \phi_{\mathbf{k}}}{\partial t} &= 0 \\[2mm] i\,|\,\mathbf{k}\,|\,\frac{\partial (A_{\mathbf{k}})_\sigma}{\partial t} + \frac{\partial^2 \phi_{\mathbf{k}}}{\partial t^2} &= 0 \end{aligned}\right\} \qquad . \qquad . \qquad (150)$$

The supplementary condition may be treated, however, in a different and more elegant manner. The transverse waves of the vector potential field give rise to electric and magnetic fields which are themselves composed of transverse waves, i.e., to solenoidal electric and magnetic fields. The longitudinal waves and the waves of the scalar potential field, on the other hand, result in no magnetic field, but in an irrotational electric field. Now it is evident from the electrodynamic equation

$$\operatorname{div} \boldsymbol{\mathcal{E}} = 4\pi\rho$$

that, on resolving the electric field into its solenoidal and irrotational components, the latter can be derived from a (non-retarded) potential determined by

$$\phi = \int \frac{\rho}{r} \, d\tau, \qquad . \qquad . \qquad . \qquad (151)$$

r specifying the distance between the point at which the potential is to be found and the element of volume $d\tau$. We conclude that the longitudinal waves and the waves of the scalar potential field can be neglected; their effect is taken into account by considering the dynamical system to possess a potential energy given by the usual Coulomb interaction formula.

As a point of interest, it is sometimes convenient to expand the field in terms of sin $(\mathbf{k.x})$ and cos $(\mathbf{k.x})$ instead of $e^{i\mathbf{k.x}}$. Thus, if $(q_{ck})_1$ and $(q_{sk})_1$ are the coefficients of cos $(\mathbf{k.x})$ and sin $(\mathbf{k.x})$ in the expansion of A_{λ_1} (and likewise for λ_2), we have

$$\left.\begin{aligned}
(q_{ck})_1 &= (A_{\mathbf{k}})_{\lambda_1} + (A_{(-\mathbf{k})})_{\lambda_1} \\
(q_{sk})_1 &= i\{(A_{\mathbf{k}})_{\lambda_1} - (A_{(-\mathbf{k})})_{\lambda_1}\}
\end{aligned}\right\} \qquad . \qquad . \qquad . \qquad (152)$$

The new momenta will then be

$$\left.\begin{aligned}
(P_{ck})_1 &= \frac{P_{(A_{\mathbf{k}})\lambda_1} + P_{(A_{-\mathbf{k}})\lambda_1}}{2} \\
(P_{sk})_1 &= \frac{P_{(A_{\mathbf{k}})\lambda_1} - P_{(A_{-\mathbf{k}})\lambda_1}}{2i}
\end{aligned}\right\}, \qquad . \qquad . \qquad (153)$$

for, with these values,

$$(P_{ck})_1 d(q_{ck})_1 + (P_{sk})_1 d(q_{sk})_1$$

$$= P_{(A_{\mathbf{k}})\lambda_1} d(A_{\mathbf{k}})_{\lambda_1} + P_{(A_{-\mathbf{k}})\lambda_1} d(A_{(-\mathbf{k})})_{\lambda_1}. \qquad . \qquad (154)$$

From (144), (152) and (153),

$$\left.\begin{aligned}
(P_{ck})_1 &= \frac{\pi^2}{c^2} (\dot{q}_{ck})_1 \\
(P_{sk})_1 &= \frac{\pi^2}{c^2} (\dot{q}_{sk})_1
\end{aligned}\right\} . \qquad . \qquad . \qquad (155)$$

The integral of the $(A_k)_r$'s in (145) is here replaced by

$$\tfrac{1}{2}\pi^2 \int \left[\mathbf{k}^2 \{ (q_{ck})_1{}^2 + (q_{sk})_1{}^2 + (q_{ck})_2{}^2 + (q_{sk})_2{}^2 \} \right.$$

$$\left. + \frac{1}{c^2} \{ (\dot{q}_{ck})_1{}^2 + (\dot{q}_{sk})_1{}^2 + (\dot{q}_{ck})_2{}^2 + (\dot{q}_{sk})_2{}^2 \} \right] d^3\mathbf{k}. \qquad (156)$$

The integration is to be taken over half the $\mathbf{k}$-space only, since the $(-\mathbf{k})$'s have been included in the above Hamiltonian.

Let us now transform to discrete variables. We realize at once from (155) and (156) that, in the absence of charges, *the individual waves may be looked upon as simple harmonic oscillators* of mass $\dfrac{\pi^2}{c^2}\, \delta^3\mathbf{k}$ ($\delta^3\mathbf{k}$ being the volume of the cell in the $\mathbf{k}$-space) and frequency ck. With the introduction of the charges, the interaction terms in the Hamiltonian can be conceived as a perturbation imposed on the oscillators by the charges. The essential consequence of our analysis is that, even in the presence of charges, the field "oscillators" and the charged bodies together constitute a system adaptable to Hamiltonian mechanics.

At the beginning of the section, we set out to examine whether variational principles and the dynamics of Hamilton embraced the theory of electromagnetism within their scope. The developments summarized in the foregoing exposition have, it will be agreed, provided a positive answer to our question. This fact, besides being of significance for general theory, has assumed an additional importance, in that it permits the transition to quantum mechanics.

9

Résumé of Variational Principles in Classical Mechanics

WE have travelled far since our introduction of the first minimal principle in physics, namely that of Hero. The reader will recollect that Maupertuis, and to a lesser degree Hero, Fermat and even Euler believed that such principles had a metaphysical import of one kind or another, which today seems rather crude. But he will also have observed how these postulates in the hands of such thinkers as Lagrange, Hamilton and Jacobi were translated into exact analytical instruments which were both highly mathematized and, at the same time, capable of solving concrete problems, particularly in astronomy. The methods thus developed disclosed in their final formulation little evidence of their metaphysical origin. Even the concept of a "minimum," which played such a momentous rôle in the work of Fermat and Maupertuis, has had to be discarded in favour of a more abstract and less anthropomorphic interpretation which is, *pari passu*, more general. Be that as it may, the source of these variational principles can still be discerned in the term "least action."

That a theory of this nature originated in the manner just described, is not surprising. Indeed, when viewed from a later period, the motives that inspired scientists in their research would often appear outmoded and vague, since the general notions of science progress with the advancement of individual theories.

This study would be incomplete if we did not stress the wide range of validity exhibited by variational principles in theoretical physics. In the preceding pages, it has already been demonstrated how they can be employed to derive the equations of optics, dynamics of particles and rigid bodies, and electromagnetism. In addition, physicists have succeeded in formulating the laws of elasticity and hydrodynamics as variational principles, and even Einstein's law of gravitation was included in this category by Hilbert, who found a scalar function $\mathfrak{H}$ for which $\delta \int \mathfrak{H} dx_0 dx_1 dx_2 dx_3 = 0$ is equivalent to Einstein's law. This function has been called the "curvature," an identification which induced Whittaker to describe Hilbert's principle in the laconic words, "gravitation simply represents a continual effort of the universe to straighten itself out."

It is commonly agreed that thermodynamics is a branch of physics which is not adaptable to the technique of variational principles. As a matter of fact, Helmholtz (1821–1894) did undertake to show that such an adaptation was possible, provided the thermic process under consideration was reversible.[22]

If one of the generalized co-ordinates q_r of a system is displaced by a distance dq_r, then

$$\text{heat taken in} = \text{work done by system} + \text{increase of internal energy,}$$

or
$$\theta dS = -Q_r dq_r + dU,$$

where θ† is the absolute temperature of the system, S the entropy, Q_r the generalized force corresponding to the co-ordinate q_r, and U the internal energy.

Helmholtz now generalizes the procedure by defining s as any function of S and introducing a new function

$$\eta = \theta \frac{dS}{ds}.$$

The energy equation thus becomes

$$\eta ds = -Q_r dq_r + dU,$$

or
$$\frac{\partial U}{\partial q_r} - \eta \frac{\partial s}{\partial q_r} = Q_r,$$

i.e.,
$$\frac{\partial F'}{\partial q_r} = Q_r,$$

† Following Helmholtz's notation, we express absolute temperature by θ, instead of by the customary T.

where F' is the function $U - \eta s$, which reduces to the free energy $U - \theta S$, when $s = S$. On taking into account the effect due to the inertia of the system, and using the Lagrange equations in the form (28), he arrives at the relation

$$\frac{d}{dt}\left(\frac{\partial T}{\partial \dot{q}_r}\right) - \frac{\partial}{\partial q_r}(T - F') = Q_r.$$

Further, from the energy equation, he obtains

$$\frac{\partial U}{\partial \eta} = \eta \frac{\partial s}{\partial \eta},$$

and therefore,

$$\frac{\partial F'}{\partial \eta} = -s.$$

If dQ^* represents the quantity of heat absorbed in an infinitesimal transformation, the ordinary definition of entropy leads to

$$\frac{dQ^*}{dt} = \theta \frac{dS}{dt} = \eta \frac{ds}{dt}$$

$$= -\eta \frac{d}{dt}\left(\frac{\partial F'}{\partial \eta}\right).$$

Finally, the total energy of the system (apart from the potential energy resulting from the forces Q_r) is given by

$$E = F' - \eta \frac{\partial F'}{\partial \eta} + T.$$

Helmholtz argues as follows: let Q_η be defined by the equation

$$Q_\eta \eta \, dt = dQ^*.$$

Suppose next that there is an additional co-ordinate corresponding to which the velocity is η and the momentum s. It is then clear that, if we assume $-F'$ to be the thermal contribution to the Lagrange function, the above equations are analogous to the usual equations of a dynamical system expressed in the form (28). Hamilton's principle still holds in dynamics with the Lagrange equations written in this manner, because, as long as the Q_r's are independent of the co-ordinates and velocities, they only necessitate an extra term $\sum_r Q_r q_r$ in the function which must be subjected to the variational condition. This is an immediate consequence of the Euler conditions for an integral to be stationary.

But any closer scrutiny will reveal that this analogy between thermodynamics and mechanics is merely formal. The relevant equations proposed by Helmholtz can certainly not serve to provide a complete description of a thermodynamical change. When we analyze the subject fully, we can at once observe that these equations are not equivalent to a variational principle. This result might have been expected from the beginning, as the very nature of thermodynamical problems seems to indicate that they do not fall under the purview of such principles.

The minimal principles discussed in this monograph are of the type which one may designate by the term "integral principles," i.e., they assert minimization of a given integral subject to certain restrictions. As minimum principles, they are not quite accurate, and have had to be amended to take the form of variational principles. Minimum conditions of an entirely different nature which, in contrast to our integral principles, can be classified as "differential principles," have occasionally been brought forward as a means of summarizing certain physical laws. The two most important are Gauss' "principle of least constraint" and Hertz's "principle of least curvature." These differential principles are in no way related to variational principles.

Gauss' principle represents an attempt at a generalization which would define the motion of a system of mass points subject to certain restrictions, if the motion of the corresponding unrestricted system were known. Suppose that at the time t the rth particle, having mass m_r, is at a point A_r, and that, if each particle were allowed to move freely, the mass point under consideration would be at B_r after a time interval dt. Let C_r denote its actual position under the restrictions imposed. The sum $\sum_r m_r (B_r C_r)^2$ was called by Gauss the "constraint," and he proved that this constraint is less for the actual motion than for any other motion compatible with the restrictions.

A similar theorem was propounded by Hertz, who considered the motion of a system subjected to kinematical restrictions, but to no dynamical forces. He defined the "length" ds of the path element, when the system is displaced so that the rth particle moves from A_r to B_r, by the equation

$$m(ds)^2 = \sum_r m_r (A_r B_r)^2,$$

where m is the mass of the system. The "angle" θ between this path element and another, in which the rth particle is displaced from A_r' to B_r', was given by means of the equation

$$mdsds' \cos \theta = \sum m_r (A_r B_r) (A_r' B_r').$$

It will readily be perceived that these definitions are generalizations of the ordinary 3-dimensional concepts of length and angle. Finally, the curvature of a path at any point is defined as the limit of the ratio of the angle between the directions at the two extremities of a path element to its length, as the element becomes infinitely small. In the case of a system free of restrictions, the curvature would be zero, or as we may say, it would travel in a "straight" path. Such a straight path is in general impossible, but Hertz showed analytically that of all path elements compatible with the restrictions, and having a given direction, the system takes that with the least curvature.

The principles of Gauss and Hertz (that of the latter generalized to systems under the influence of dynamical forces) are perhaps more conveniently stated in the form that, if our attention is confined to motions kinematically possible under the forementioned restrictions, the value of the expression

$$\sum_r m_r \left\{ \left(\ddot{x}_r - \frac{X_r}{m_r} \right)^2 + \left(\ddot{y}_r - \frac{Y_r}{m_r} \right)^2 + \left(\ddot{z}_r - \frac{Z_r}{m_r} \right)^2 \right\}$$

is a minimum; the velocities of the elements of the system are, as usual, supposed to be fixed. This statement of the principle is mathematically preferable to those of Gauss and Hertz, because it enables us to determine the motion of the system under general conditions, without prior knowledge of how the system would move when not subjected to the restrictions. Certain physicists may, on the other hand, deem it aesthetically inferior, since the forces have to be introduced.

These theorems are easily interrelated and their proofs are straightforward. We shall not, however, reproduce them here, as they are rather outside the sphere of this work.

Throughout the preceding exposition we have seen how action principles and the Hamilton-Jacobi equation are intimately connected with pre-quantum mechanics, both classical and relativistic. It will now be our purpose to investigate the extent to which these concepts have been applied, not so much in quantum mechanics itself, but in the theories leading up to the epoch-making discoveries which today constitute the basis for the mechanics of elementary particles.

§ 10

Relation between Variational Principles and the Older Form of Quantum Theory

By common consent, quantum mechanics is the most revolutionary and comprehensive achievement in modern theoretical physics. However, the great changes, systematized into a consistent and coherent theory between 1925 and 1928, came about only gradually during the first quarter of the century. Throughout this period, hypothesis after hypothesis, each bringing us nearer to the imposing mathematico-physical edifice of today, had to be abandoned owing to logical inconsistencies and failure to produce agreement with observation and experiment. In this section, we shall endeavour to outline the part which the concept of action has occupied in these developments.

The early stages of quantum theory are well-known and warrant but brief recapitulation here. In 1900, Planck postulated that light could not be radiated continuously, but only in whole multiples of $h\nu$, where ν is the frequency and h a *new universal constant known as "Planck's constant,"* its most recently computed value being $6 \cdot 624.10^{-27}$ erg sec. The introduction of this hypothesis was rendered necessary in order to explain the spectral distribution of black body radiation; the black body was supposed to contain simple harmonic oscillators whose energy could only be integral multiples of $h\nu$. It cannot be emphasized sufficiently that Planck's discovery signifies a decisive breach with classical physics and is incompatible with the main body of nineteenth-century mechanics.

In 1905, Einstein pointed out that the experimental data relevant to the photo-electric effect were easily interpreted if it

97

were assumed that light travels through space in "bundles" of energy $h\nu$; furthermore, the value of h calculated from measurements of this kind coincided, within the margin of experimental error, with the value found by Planck.

The third phase in the progress of quantum theory was initiated by Bohr.[23] It had already been suggested by Rutherford, on the basis of scattering experiments, that atoms are composed of electrons revolving round a central positively charged nucleus. But this proposal gave rise to the difficulty that, according to the classical theory of electrodynamics, such an electron would continually lose energy by radiation and therefore eventually fall into the nucleus. In consequence, Bohr put forward the hypothesis that electrons do not radiate energy continuously: they can revolve in certain discrete orbits without radiating, and, when "jumping" from one state to another, will emit a quantum of light of frequency given by

$$h\nu = E_2 - E_1, \qquad . \qquad . \qquad . \qquad (157)$$

where E_2 and E_1 are the energies of the first and second orbit respectively. This condition does not, however, give us any information as to which paths the electrons may occupy. To define these states, Bohr, concentrating upon circular motion in a hydrogen-like atom, provisionally assumed that when an electron "drops" into an orbit from infinity, the frequency radiated is half the frequency of the electron, i.e., the average of the frequencies before and after the jump. Thus, if E is the energy of the orbit,

$$E = -\tfrac{1}{2}nh\nu, \qquad . \qquad . \qquad . \qquad (158)$$

n being a positive integer. From the ordinary formula for circular motion we obtain

$$4\pi^2 mv^2 r = \frac{Ze^2}{r^2}. \qquad . \qquad . \qquad . \qquad (159)$$

(m = mass of electron, e = charge on electron, Z = number of charges on nucleus, r = radius of orbit).

Now

$$E = \tfrac{1}{2} \cdot 4\pi^2 mv^2 r^2 - \frac{Ze^2}{r},$$

or, from (159),

$$E = -\tfrac{1}{2}\frac{Ze^2}{r} \qquad . \qquad . \qquad . \qquad (160)$$

$$= -2\pi^2 mv^2 r^2 \qquad . \qquad . \qquad . \qquad (161)$$

$$= -\pi p_\theta \nu, \qquad . \qquad . \qquad . \qquad (162)$$

where p_θ is the angular momentum given by

$$p_\theta = 2\pi m v r^2. \qquad . \qquad . \qquad . \qquad (163)$$

Combining (158) with (162), we derive the most important relation

$$p_\theta = \frac{nh}{2\pi}. \qquad . \qquad . \qquad . \qquad (164)$$

In words: *the angular momentum must be an integral multiple of $h/2\pi$.* Eliminating r and v from eqs. (160), (161) and (163), we find that the energies of the possible orbits are determined by

$$E = -\frac{2\pi^2 m Z^2 e^4}{n^2 h^2}. \qquad . \qquad . \qquad (165)$$

From (157) and (165), Bohr finally arrived at the following equation for the possible frequencies of the hydrogen spectrum, namely,

$$\nu = \frac{2\pi^2 m e^4}{h^3}\left(\frac{1}{n_1{}^2} - \frac{1}{n_2{}^2}\right). \qquad . \qquad . \qquad (166)$$

This equation agreed very closely with the experimental data of Balmer and others, thereby confirming Bohr's theory. At the same time, it provided a remarkable manifestation of the universality of the constant h.

Bohr then scrutinized further his main assumption (158). In place of this hypothesis, he supposed that, for the nth orbit,

$$E = -f(n)\,h\nu; \qquad . \qquad . \qquad . \qquad (167)$$

f is dimensionally a purely numerical function. He found for the Balmer formula, on pursuing the same course as before,

$$\nu = \frac{\pi^2 m e^4}{2h^3}\left[\frac{1}{\{f(n_1)\}^2} - \frac{1}{\{f(n_2)\}^2}\right] \qquad . \qquad . \qquad (168)$$

In order that the last equation should comply with the empirical Balmer formula, it is evident that

$$f(n) = cn. \qquad . \qquad . \qquad . \qquad (169)$$

To calculate c, Bohr employed the following ingenious device. The frequency of the light emitted when the electron jumps from the nth orbit to the $(n-1)^{\text{th}}$ is, using eqs. (168) and (169),

$$\nu = \frac{\pi^2 m e^4}{2c^2 h^3} \cdot \frac{2n-1}{n^2(n-1)^2}. \qquad . \qquad . \qquad (170)$$

Also, an application of eqs. (158) and (165), modified appropriately to include the constant c, gives for the frequency of the nth orbit

$$\nu = \frac{\pi^2 m e^4}{2c^3 h^3 n^3}. \qquad . \qquad . \qquad . \qquad (171)$$

Bohr then hypothesized that, if n is large, the frequencies of the orbit and of the emitted radiation should be the same, because the electron would then radiate according to classical theory. Comparing (170) with (171), we get

$$c = \tfrac{1}{2},$$

thus substantiating eq. (158).

Bohr's theory made it possible to apply the quantum conditions only to the very restricted case of an electron moving in a circular orbit about a centre of force. An extension to more general cases was evolved a few years later by Sommerfeld,[24, 25] who took as his starting-point an idea originally put forward by Planck. Considering a 1-dimensional simple harmonic oscillator, Planck wished to compute the probability that the system is in any particular quantized state. Liouville's theorem had established that, in classical mechanics, equal elements of the phase space, i.e., the 2-dimensional pq-space, are equally probable. Planck then examined the graphs of p against q for his quantized simple harmonic oscillators. The equation of a characteristic curve is

$$\frac{p^2}{2m} + 2\pi^2 m \nu^2 q^2 = E, \qquad . \qquad . \qquad (172)$$

which represents an ellipse of area E/ν. Since the permissible states of the oscillator have energy $nh\nu$, it follows that the area of the nth ellipse is nh, or, that the area between any two consecutive ellipses is h. This result, together with Liouville's theorem, leads to the conclusion that *all states are equally probable*. Planck accordingly suggested as a general law pertaining to all 1-dimensional systems, that only certain states were possible, and that the area in phase space between the graphs of two such consecutive states was equal to h. It therefore appeared that phase space was not infinitely divisible, but was built up of elements of area h, and because such an area has dimensions pq, Planck called h the "elementary quantum of action."

Sommerfeld drew attention to the circumstance that Planck's hypothesis held in connexion with Bohr's circular orbits as well as with simple harmonic oscillators. In the former case, the graph of the constant angular momentum against angular position is a horizontal straight line and not a closed curve. He argued, however, that artificial boundaries could be constructed by means of ordinates at a distance of 2π from each other, for an increase of 2π in an angular co-ordinate is tantamount to a return to the initial position. As the angular momentum has to be a multiple of $h/2\pi$, the area between two consecutive curves is h. Proceeding to the general case, Sommerfeld expressed Planck's hypothesis in the form

$$\left(\oint p\,dq\right)_n - \left(\oint p\,dq\right)_{n-1} = h,$$

the suffix n referring to the nth orbit. The sign $\oint$ signifies integration round a complete curve, and the integral is known as a "phase integral." He further assumed that

$$\left(\oint p\,dq\right)_0 = 0,$$

and hence found

$$\oint p\,dq = nh. \qquad . \qquad . \qquad (173)$$

Our familiar action integral is here introduced into quantum theory!

Sommerfeld next treated the hydrogen atom on the basis of this theory, without confining himself to circular orbits. He first attempted to quantize the angular momentum alone, that is, to use Bohr's angular momentum condition, but discovered that a continuum of energies would result if elliptic orbits were taken into account. A second and more fruitful method was to quantize both the angular and the radial momenta, the conditions being

$$\left. \begin{aligned} \oint p_\theta\,d\theta &= n_1 h,\ \text{i.e.,}\ p_\theta = \frac{n_1 h}{2\pi} \\[2mm] \oint p_r\,dr &= n_2 h \end{aligned} \right\} \qquad . \qquad . \qquad (174)$$

It is usual to call the integers n_1 and n_2 the "azimuthal" and "radial" quantum numbers respectively.

The energy of the orbits was found to be

$$E = -\frac{2\pi^2 me^4}{h^2(n_1 + n_2)^2}, \qquad . \qquad . \qquad . \qquad (175)$$

and thus the Balmer formula is once more satisfied. It will be noticed that the energy depends only on the sum of the two quantum numbers, and not on each separately. In a later communication, Sommerfeld presented a solution of the problem of the energy levels in the hydrogen-like atom from the viewpoint of relativity mechanics and showed that, in this case, the energy does depend slightly on the individual quantum numbers; that is to say, each line in a spectrum of a hydrogen-like atom is split up into a "fine-structure." This result agreed qualitatively and quantitatively with experiment, especially in regard to the spectrum of ionized helium, where the splitting is larger and therefore more easily measurable than that of hydrogen. Sommerfeld's explanation of fine-structure was rightly appraised as one of the triumphs of quantum theory. We shall not, however, enter into his calculations here, because the formulae will be derived later by less tedious operations.

The historical priority in the discovery of the quantization of the phase integrals belongs not to Sommerfeld, but to W. Wilson,[26] who had advanced the same hypothesis a few months earlier. Wilson supposed that co-ordinates had been chosen in such a way that the expression for the kinetic energy did not contain product terms and could thus be written in the form

$$T = \sum_r \tfrac{1}{2} C_r \dot{q}_r^2, \qquad . \qquad . \qquad . \qquad (176)$$

and he defined T_r by means of the equation

$$T_r = \tfrac{1}{2} C_r \dot{q}_r^2. \qquad . \qquad . \qquad . \qquad (177)$$

After stipulating that the motion of the system had to be periodic in each of the co-ordinates, he set as his quantum conditions

$$2\oint T_r \, dt = n_r h,$$

i.e.,
$$\oint p_r \, dq_r = n_r h.$$

The theory of Sommerfeld and Wilson suffered conspicuously from the defect that it was not able to suggest any rules for the choice of co-ordinates to which the quantum conditions were to be applied. For the special case of the Kepler ellipse, the selection of r and θ seems obvious, but such a distinct choice does not exist in more complicated systems. A first indication as to how to approach such problems was given by Epstein in a paper on the Stark effect.[27] Epstein carried over from astronomy a method which had been much employed for solving the Hamilton-Jacobi equation and which is known as the method of "separation of the variables." The procedure is best illustrated by means of a concrete example and we shall, for this purpose, compute the energy levels of a hydrogen-like atom according to relativity theory. The energy is expressed by the formula

$$\frac{1}{c^2}\left(E + mc^2 + \frac{Ze^2}{r}\right)^2 = m^2c^2 + p_r{}^2 + \frac{1}{r^2}p_\theta{}^2.$$

The Hamilton-Jacobi equation therefore reads

$$m^2c^2 + \left(\frac{\partial S}{\partial r}\right)^2 + \frac{1}{r^2}\left(\frac{\partial S}{\partial \theta}\right)^2 = \frac{1}{c^2}\left(E + mc^2 + \frac{Ze^2}{r}\right)^2. \dagger \qquad (178)$$

We may then put

$$S = S_1(\theta) + S_2(r), \qquad . \qquad . \qquad . \qquad (179)$$

where S_1 and S_2 satisfy respectively the equations

$$\frac{dS_1}{d\theta} = \alpha_1, \qquad . \qquad . \qquad . \qquad . \qquad (180)$$

$$m^2c^2 + \left(\frac{dS_2}{dr}\right)^2 + \frac{\alpha_1{}^2}{r^2} = \frac{1}{c^2}\left(E + mc^2 + \frac{Ze^2}{r}\right)^2, \qquad (181)$$

α_1 being an arbitrary constant. The integration of (179) has hereby been reduced to the solution of two ordinary differential equations, and furthermore, the function $S_1 + S_2$ is of the type required, as it contains one arbitrary constant α_1 apart from the additive constant resulting from the integration of (180) and (181).

From eq. (180),

$$S_1 = \alpha_1\theta + c_1 \qquad . \qquad . \qquad . \qquad (182)$$

and, from eq. (181),

$$S_2 = \int\left\{\sqrt{-\frac{1}{r^2}\left(\alpha_1{}^2 - \frac{Z^2e^4}{c^2}\right) + \frac{2Ze^2}{rc^2}(E + mc^2) + \left(\frac{E^2}{c^2} + 2Em\right)}\right\}dr.$$

$$(183)$$

† When dealing with quantum theory, we shall always operate with eq. (62) and not with (63), for it is the former that is related to the action function.

We are now in a position to introduce the quantum conditions. To this end, we need merely revert to eqs. (67), from which it follows that

$$\oint p_\theta \, d\theta = \int_0^{2\pi} \frac{dS_1}{d\theta} \, d\theta = n_1 h \qquad . \qquad . \qquad (184)$$

and

$$\oint p_r \, dr = \oint \frac{dS_2}{dr} \, dr = n_2 h. \qquad . \qquad . \qquad (185)$$

Eqs. (182) and (184) combined lead to

$$\alpha_1 = \frac{n_1 h}{2\pi} . \qquad . \qquad . \qquad . \qquad (186)$$

The evaluation of the phase integral (185) demands further investigation. The radius r varies between two extreme limits r_0 and r_1 with the rotation of the particle, and we have to integrate over the range $r_0 \to r_1 \to r_0$. When r has one of its extreme values, p_r and therefore the integrand in eq. (183) become zero. Hence, r_0 and r_1 are the roots of the integrand, which must be real between these two values. The sign of the radical has to be chosen so that $p_r dq_r$ is positive. On interpreting the phase integral in this way, its value, most effectively computed with the aid of contour integration, is found to be

$$2\pi \left\{ - \sqrt{\alpha_1{}^2 - \frac{Z^2 e^4}{c^2}} + \frac{Z e^2 (E + mc^2)}{c^2 \sqrt{-\dfrac{E^2}{c^2} - 2Em}} \right\}. \qquad . \qquad (187)$$

If we set the expression (187) equal to $n_2 h$ and substitute $\alpha_1 = n_1 h / 2\pi$, we obtain for E,

$$E = mc^2 \left[\left\{ 1 + \frac{Z^2 \alpha^2}{(n_2 + \sqrt{n_1{}^2 - \alpha^2 Z^2})^2} \right\}^{-\frac{1}{2}} - 1 \right], \quad (188)$$

where α is the dimensionless constant $2\pi e^2 / hc$, known as the "fine-structure constant." Eq. (188) is Sommerfeld's fine-structure formula. When c approaches infinity, (188) tends to the non-relativistic equation (175).

The consummation and simplicity of this method will have impressed the reader, as it leads directly to the energy levels without intermediate calculations of the orbits. In relativity mechanics, these calculations would be rather complicated.

More important, it provides an answer to the question regarding the choice of co-ordinates to which the quantum conditions are to be applied; we must single out those co-ordinates that allow the Hamilton-Jacobi equation to be separated. The uniqueness of these co-ordinates will be discussed later. A procedure similar to that adopted in this example can always be followed if the Hamilton-Jacobi equation is separable, with the result that the n phase integrals will serve to determine the $(n - 1)$ constants α_r and the constant E. The energy levels will thus depend on the phase integrals alone, whereas, if we were to select any other co-ordinates, this criterion need not necessarily hold. In fact, such co-ordinates might not vibrate between fixed limits, and it would then be impossible to define phase integrals.

Another attack on the problem under review was made by Schwarzschild independently of Epstein. Before outlining Schwarzschild's contribution, it is advisable, however, to refer briefly to the Staude-Stäckel theory of conditionally periodic motion. Suppose we have a system with n generalized co-ordinates, and let us further assume that the Hamilton-Jacobi equation is separable, so that

$$S = \sum_r S_r(q_r), \qquad . \qquad . \qquad . \qquad (189)$$

where the S_r's are functions of q_r, $(n - 1)$ constants α_s and the constant E. The S_r's will be of the form

$$\int \sqrt{f_r(q_r, \alpha_1, \ldots, \alpha_{n-1})} \, dq_r.$$

Each f_r will have two real roots q_{r0} and q_{r1}, between which it is positive. Once S is separated in this manner, eqs. (66a and b) become

$$
\left.
\begin{aligned}
(a) \qquad \beta_s &= \frac{\partial S}{\partial \alpha_s} = \sum_r \int \frac{\phi_{rs}(q_r)}{\sqrt{f_r(q_r)}} \, dq_r \\
(b) \quad \beta_0 + t &= \frac{\partial S}{\partial E} = \sum_r \int \frac{\phi_{r0}(q_r)}{\sqrt{f_r(q_r)}} \, dq_r
\end{aligned}
\right\} , \qquad . \qquad (190)
$$

denoting $\dfrac{1}{2}\dfrac{\partial f_r}{\partial \alpha_s}$ by ϕ_{rs} and $\dfrac{\partial f_r}{\partial E}$ by ϕ_{ro}. The β's are constants by Jacobi's theorem. Now define ω_{rs} by $\oint \dfrac{\phi_{rs}(q_r)}{\sqrt{f_r(q_r)}} \, dq_r$, the phase integral to be taken in the customary way. We thereupon notice that, if q_r undergoes one complete oscillation, the other q's remaining constant, β_s will increase by ω_{rs} and $\beta_0 + t$ by ω_{r0}. Conversely, when β_s increases by ω_{rs} and $\beta_0 + t$ by ω_{r0}, q_r will pass through one complete oscillation, while the other q's remain constant. As the next step, define the new variables w_r by the equations

$$(a) \qquad \beta_0 + t = \sum_r w_r \omega_{r0} \; \Big\rbrace$$

$$(b) \qquad \beta_s = \sum_r w_r \omega_{rs} \; \Big\rbrace \qquad . \qquad . \qquad . \quad (191)$$

Obviously, an increase of unity in w_r, with the other w's kept constant, will result in an increase of ω_{r0} in $\beta_0 + t$ and an increase of ω_{rs} in β_s; q_r will thereby undergo one complete oscillation while the other q's remain constant. Hence the q's are periodic functions of unit period in the w's, which, as it is clear from eqs. (191), increase uniformly with the time t. Accordingly, we can write

$$w_r = \omega_r t + \text{const.} \qquad . \qquad . \qquad . \quad (192)$$

The co-ordinates may be expanded in a Fourier series, so that

$$q_r = \sum e^{i(\sum_r n_r w_r)}, \qquad . \qquad . \qquad . \quad (193)$$

the n's taking integral values. Therefore

$$q_r = \sum e^{i(\sum_r n_r \omega_r)t}. \qquad . \qquad . \qquad . \quad (194)$$

Such motion is designated "conditionally periodic."

If one or more of the co-ordinates q_r is an angle which increases continuously with the motion, the derivative dS_r/dq_r will be constant, and the phase integral is to be interpreted as the integral over the range 0 to 2π.

After this digression, we may return to our immediate object. The co-ordinates q_r and the momenta p_r are functions of the β's, α's and E, and therefore of the w's, α's and E; during the motion, the α's and E are constant. Schwarzschild[28] commenced with the supposition that the α's and E had been replaced by constant quantities J_r conjugate to the w_r's. Hamilton's canonical equations in terms of these co-ordinates read

$$\frac{dw_r}{dt} = \frac{\partial H}{\partial J_r}; \qquad \frac{dJ_r}{dt} = -\frac{\partial H}{\partial w_r}.$$

The second of these equations together with the constancy of the J_r's permits us to infer that H is a function of the J_r's only, whereas the first confirms that w_r increases uniformly with time, the ω_r's of eq. (192) being given by the formula

$$\omega_r = \frac{\partial H}{\partial J_r}. \qquad . \qquad . \qquad . \qquad (195)$$

In virtue of the fact that the q's and p's are periodic of period unity in the w's, these latter variables are plainly dimensionless, and since the product of a co-ordinate and its conjugate momentum has the dimensions of action, the J's have also the dimensions of action. Hence Schwarzschild calls the J's "action variables," as opposed to the w's which, in consequence of their periodicity property, are termed "angle variables." On the strength of these remarks, the manner in which Sommerfeld's quantum conditions are to be applied presents itself at once. For it is obvious that the phase integral $\int_0^1 J_r \, dw_r$ is equal to J_r; and the desired method of quantization, as proposed by Schwarzschild, is that the J's must be integral multiples of h.

Schwarzschild treated the theory of the Stark effect on the basis of these quantum conditions, and thereby actually reached results fully equivalent to those of Epstein. This relationship was made rather more explicit in a later paper by Epstein,[29] who showed that his quantum conditions were tantamount to Schwarzschild's. To wit, denote Epstein's phase integrals provisionally by J_r. It follows that

$$\omega_{rs} = \frac{\partial J}{\partial \alpha_s}, \qquad . \qquad . \qquad . \qquad (196)$$

where ω_{rs} denotes $\oint \frac{\phi_{rs}(q_r)}{\sqrt{f_r(q_r)}} \, dq_r$ as before. Then,

$$\left.\begin{array}{l} \beta_s = \dfrac{\partial S}{\partial \alpha_s} = \sum_r \dfrac{\partial S}{\partial J_r} \omega_{rs} \\[2ex] \beta_0 + t = \dfrac{\partial S}{\partial E} = \sum_r \dfrac{\partial S}{\partial J_r} \omega_{r0} \end{array}\right\}, \qquad . \qquad . \qquad (197)$$

whence, by comparison with eqs. (191), we deduce that

$$w_r = \frac{\partial S}{\partial J_r}. \qquad . \qquad . \qquad . \qquad (198)$$

Eq. (198), combined with the equation $p_r = \partial S / \partial q_r$, enables us to regard the sets of variables w_r and J_r as canonically conjugate, and further, J_r, which is a function of the α's and E only, does not vary with time. Co-ordinating all these results, we can conclude that the J's are our required action variables and that Epstein's quantum conditions are identical with those of Schwarzschild.

It remains to examine the question whether the Hamilton-Jacobi equation might not be separable in more than one system of generalized co-ordinates. If this were the case, Epstein's method would not determine unambiguously the quantum

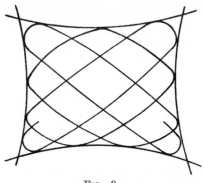

Fig. 9

conditions. We have seen, earlier on in this section, that each co-ordinate q_r oscillates between fixed limits, and the motion of such a system in two dimensions would thus be depicted graphically as in Fig. 9.

As long as the frequencies are incommensurable, the path will not repeat itself and the limits of the co-ordinates will form an envelope of the curve. In consequence, the surfaces $q_r = $ const. are uniquely determined. It results therefrom that the relation between any two sets of co-ordinates, in which the Hamilton-Jacobi equation is separable, may be written

$$Q_r = f_r(q_r),$$

and with this transformation of co-ordinates, $p_r dq_r$ is invariant, whence the quantum conditions are unaltered.

In the case of "degenerate systems," i.e., systems in which two or more frequencies $\omega_i, \omega_j, \ldots$ are equal—a feature which pertains, for instance, to the non-relativistic Kepler ellipse—the above observations do not hold for the co-ordinates corresponding to these frequencies. However, in such situations, eqs. (195) reveal that

$$\frac{\partial H}{\partial J_i} = \frac{\partial H}{\partial J_j} = \cdots$$

and that H therefore depends only on $J_i + J_j + \ldots$. Because any change of co-ordinates which involves q_i, q_j, ... alone does not affect the value of $\sum\limits_{i,j,\ldots} p_i dq_i$, it can be asserted that the energy levels are independent of our choice of co-ordinates, although it should be borne in mind that in any new co-ordinate system different orbits would be allowed by the quantum conditions.

This conclusion was, at the time, interpreted in several alternative ways. It could have been contended that the actual orbits were unimportant, since the energy levels, which are the observed data, were uniquely determined by the quantum conditions. An approach of this kind, more consistently applied, later realized far-reaching success in the hands of Heisenberg and others. Another interpretation was to postulate that the quantum condition is imposed exclusively on the sum $J_i + J_j + \ldots$ and not on the individual phase integrals,—a suggestion that was first proposed by Schwarzschild in his original paper. Perhaps the most popular theory of systems with two or more equal frequencies consisted in the assumption that they were approximations to non-degenerate systems. For, in practice, there are always relativistic effects, electric and magnetic fields, etc., and completely accurate degeneracy does probably not exist. The whole problem concerning the quantization of degenerate systems on the basis of the Sommerfeld-Epstein-Schwarzschild conditions is today, of course, irrelevant, as these conditions have been superseded by more modern theories.

Two cardinal principles of the older quantum theory may here be shortly outlined, namely the "correspondence principle" of Bohr and the "adiabatic hypothesis" of Ehrenfest. We have already mentioned how Bohr, in his original paper on the quantization of electronic orbits, utilized the criterion that the frequency of the emitted radiation had to tend to the classical frequency as the size of the orbits approached infinity. It will now be shown that this is a general result of the Bohr-Sommerfeld quantum hypothesis. From eq. (194), one observes that, classically, the electron emits radiation whose Fourier components have frequencies of the type $\sum\limits_{r} n_r \omega_r$. In quantum theory, on the other hand, the frequencies are given by

$$\nu = \frac{E_2 - E_1}{h},$$

where E_2 and E_1 are two possible energy levels. Thus, for large quantum numbers, we may write,

$$\nu = \frac{1}{h} \sum_r \frac{\partial H}{\partial J_r} \, \delta J_r \, .$$

As the phase integrals are whole multiples of h,

$$\nu = \sum_r n_r \frac{\partial H}{\partial J_r},$$

which equation together with (195) proves that, in the limit, the classical and quantum frequencies become the same.

It often occurs that, in classical theory, some of the Fourier components of the radiation are absent. In such cases, we have to assume that the corresponding quantum "jumps" are forbidden. One of the most frequently encountered instances of these "selection rules" is the restriction that, when a particle moves under the influence of a central force, the azimuthal quantum number can only change by ± 1. To verify this rule, we notice that r vibrates with a frequency ω_r, while the angular motion may be regarded as a uniform rotation of frequency ω_θ, upon which is superimposed a periodic vibration of frequency ω_r. On setting $x = r \cos \theta$, $y = r \sin \theta$, it is seen without difficulty that x and y may be expanded in the form

$$\sum_{n=-\infty}^{\infty} \{ A_n e^{i(n\omega_r + \omega_\theta)} + B_n e^{i(n\omega_r - \omega_\theta)} \}.$$

In this manner, we arrive at the conclusion that changes in the azimuthal quantum number must be restricted to ± 1, if the quantum frequencies are to agree with the classical frequencies for large quantum numbers. This selection rule has greatly facilitated the interpretation of atomic spectra and the study of the electronic configurations in atoms.

The second of these fundamental principles in quantum theory, Ehrenfest's adiabatic hypothesis,[30] represented the most convincing attempt to justify Sommerfeld's quantum conditions theoretically. Ehrenfest had previously been working (in thermodynamics) with "adiabatic invariants," i.e., quantities which remain unchanged when the conditions in a dynamical system are varied *infinitely slowly* and *unsystematically*. His proposal maintained that the magnitudes which were to be quantized had to be adiabatically invariant, because, on changing the conditions of the system, these quantities would have to remain integral multiples of h. It was later demonstrated by Burgers[31] that the phase integrals are adiabatic invariants.

The proof employs the transformation function

$$S^* = S - \sum_r J_r w_r . \qquad \qquad (199)$$

In accordance with the theory of contact transformations, we will then have, if we express S^* as a function of the q's and w's,

$$p_r = \frac{\partial S^*}{\partial q_r}; \qquad J_r = - \frac{\partial S^*}{\partial w_r}. \qquad (200)$$

When w_r is augmented by unity, q_r will undergo a complete oscillation, so that S increases by J_r and S^* remains unaltered. The new function S^* is therefore, like the p's and q's, a periodic function of period unity in the w's. Proceeding with Burgers' argument, let

$$H = H(q_r, p_r, c),$$

where c is a parameter which varies slowly with the time. S^* will naturally involve c; the dependence of the w's and J's on the p's and q's will also change with time. From eqs. (83b) we get

$$\frac{dJ_r}{dt} = - \frac{\partial \bar{H}}{\partial w_r}.$$

As before,

$$\bar{H} = H + \frac{\partial S^*}{\partial t},$$

so that

$$\bar{H} = H + \frac{\partial S^*}{\partial c}\frac{dc}{dt}.$$

Hence

$$\frac{dJ_r}{dt} = - \frac{\partial}{\partial w_r}\left(\frac{\partial S^*}{\partial c}\frac{dc}{dt}\right), \quad \left(\text{since } \frac{\partial H}{\partial w_r} = 0\right),$$

and the increase in J_r over a finite time interval is

$$-\int \frac{\partial}{\partial w_r}\left(\frac{\partial S^*}{\partial c}\frac{dc}{dt}\right) dt.$$

Now $\partial S^*/\partial c$ can be expanded as a Fourier series analogous to (193). The expression $\dfrac{\partial}{\partial w_r}\left(\dfrac{\partial S^*}{\partial c}\right)$ will consequently be such a Fourier series without a constant term and will thus vary with time according to (194). Provided none of the ω's are equal, this series, too, does not contain a constant term, and the integral

$$\int \frac{\partial}{\partial w_r}\left(\frac{\partial S^*}{\partial c}\right) dt$$

is bounded. As long as c varies unsystematically, the integral

$$\int \frac{\partial}{\partial w_r} \left(\frac{\partial S^*}{\partial c} \frac{dc}{dt} \right) dt$$

will therefore tend to zero with dc/dt, even when the time over which the integral is taken increases correspondingly in such a way as to keep the change in c finite. It follows that J_r does remain constant if the system is altered infinitely slowly. We must, however, reiterate that this is only valid subject to the condition that the ω's are all unequal, i.e., that the system is non-degenerate and does not pass through a degenerate state during the process. The method of reasoning by which this theorem was used to justify Sommerfeld's quantum conditions was to imagine the system gradually modified, until its motion had become a combination of simple harmonic oscillations at right angles. In view of the fact that the phase integrals do not change during the transformation and are, according to Planck's original hypothesis, integral multiples of h at the end of the process, they must have been integral multiples of h at the beginning. It was assumed that the system was non-degenerate.

Whoever reflects upon the mathematical foundations of the Bohr-Sommerfeld quantum theory will no doubt be struck by the unexpected reappearance of the classical concept of action in an entirely new context. We had already explained how Newton's laws of motion could be formulated as the principle of the conservation of energy and the principle of least action, and how the later development of the action integral as a function of the co-ordinates culminated in the Hamilton-Jacobi equation. It now emerged that, in those systems where definite quantities of action could be associated with each state of motion, viz., in conditionally periodic systems, the action was subject to a restriction totally foreign to classical dynamics: it could only occur in integral multiples of a fixed quantum of action h.

There are two aspects from which the remarkable relationship between the action principle and quantum theory can be viewed. In the first place, when we carry out calculations with regard to the effect of the quantization of action on the energy levels, we have at our disposal an almost ready-made instrument in the Hamilton-Jacobi theory, which bases the solution of dynamical problems on the action function S, the function to which the quantum conditions are to be applied. Furthermore, since Pythagoras, the notions of both minima and whole numbers have fascinated certain metaphysically inclined scientists.   It is consequently understandable that the attention of these physicists was widely aroused by the presence of copious experimental evidence which showed, on the one hand, that the laws of mechanics can be summarized in the principle of least action and, on the other hand, that this same action function is to be subjected to integer restrictions according to quantum theory.   We shall, however, defer the epistemological evaluation of action principles to the final chapter of this monograph.

It will not have escaped the reader's attention that one significant distinction exists between the action integrals in this and in previous sections.   In classical theory, the expression encountered is always $\int \sum_r p_r dq_r$, whereas in quantum theory it is simply $\int p_r dq_r$.   This is directly connected with the fact that, in quantum theory, the action function S has of necessity to be split up into partial action functions $S_r(q_r)$, while in the Hamilton-Jacobi theory the separation of variables was merely a means of solving the equation.   The question at once arises as to whether the Hamilton-Jacobi equation can be separated in general, and the answer is decidedly in the negative.   Indeed, for the elementary case of the three-body problem, the equation cannot be separated.   As a result, it is impossible to apply Sommerfeld's conditions to determine the energy levels of the poly-electronic atoms.   We need hardly recount the diverse proposals which were proffered to subject such atoms in part to the quantum conditions, especially with respect to the angular momentum, but these efforts were not at all satisfactory and numerous heuristic *ad hoc* hypotheses were successively advanced and abandoned.   Probably the greatest defect of the Bohr-Sommerfeld theory lay, however, in its very nature; it computed the orbits on classical lines, only to reject most of them at a later stage.   Physicists were completely in the dark concerning the mechanism of transition from one orbit to another and the behaviour of the electron during this transition.   In consequence, the older version of quantum theory, its initial spectacular successes notwithstanding, ultimately suffered the fate of so many scientific theories and had to give way to more modern developments.

Postscript

In connexion with Sommerfeld's fine-structure formula (188) Professor E. Schrödinger writes, *inter alia*, in a letter dated 29th February, 1956: ". . . you are naturally aware of the fact that

Sommerfeld's derivation of the fine-structure formula provides only fortuitously the result demanded by experiment. One may notice then from this particular example that the newer form of quantum theory (i.e., quantum mechanics) is by no means such an inevitable continuation of the older theory as is commonly supposed. Admittedly the Schrödinger theory, relativistically framed (without spin), gives a *formal* expression of the fine-structure formula of Sommerfeld, but it is *incorrect* owing to the appearance of half-integers instead of integers. My paper in which this is shown has . . . never been published; it was withdrawn by me and replaced by the non-relativistic treatment. . . . The computation [by the relativistic method] is far too little known. It shows in one respect how *necessary* Dirac's improvement was, and on the other hand that it is wrong to assume that the older form of quantum theory is 'broadly' in accordance with the newer form."

We shall derive Schrödinger's non-relativistic wave equation in the next chapter; for a detailed derivation of its inadequate relativistic counterpart, as well as for an account of the correct treatment finally developed by Dirac, the reader is referred to textbooks on quantum mechanics. Here we shall be content to illustrate Professor Schrödinger's remarks by stating the relevant fine-structure formulae.

In attempting a relativistic description of hydrogen-like atoms it is natural to take as point of departure the equation preceding eq. (178), extended to three dimensions, which is, as will be explained in §11, converted into a differential equation by making the substitution $\mathbf{p} = (h/2\pi i)$ grad. The resulting wave equation would describe a hydrogen atom were it not for the fact that the particle so characterized has no spin, and thus cannot be an electron. Be that as it may, an analysis of this equation leads, as Schrödinger found, to

$$E + mc^2 = mc^2 \left\{ 1 + \frac{Z^2\alpha^2}{(n - l - \frac{1}{2} + \sqrt{(l + \frac{1}{2})^2 - \alpha^2 Z^2})^2} \right\}^{-\frac{1}{2}},$$

$$n = 1, 2, \ldots; \quad l = 0, 1, \ldots, n - 1.$$

On expanding this expression in powers of α^2, we can easily verify that the total spread of the fine-structure levels for a given n is, to terms of α^4,

$$\frac{mc^2 Z^4 \alpha^4}{n^3} \frac{n - 1}{n - \frac{1}{2}}.$$

This quantity is much larger than the value observed experimentally in the spectrum of hydrogen.

Dirac's relativistic theory for a particle with spin gives, on the other hand, a result equivalent to the formula (188) derived by Sommerfeld, namely

$$E + mc^2 = mc^2 \left\{ 1 + \frac{Z^2\alpha^2}{(n - |k| + \sqrt{k^2 - \alpha^2 Z^2})^2} \right\}^{-\frac{1}{2}},$$

$$n = 1,2, \ldots; \quad |k| = 1,2, \ldots, n.$$

The total spread in energy of the fine-structure levels for a given n now becomes

$$\frac{mc^2 Z^4 \alpha^4}{n^3} \frac{n - 1}{2n}.$$

This is substantially less than the Schrödinger value, and is in agreement with experiment.

To sum up: a valid theory of fine-structure must include both wave mechanics and spin. Sommerfeld's explanation was successful because the neglect of wave mechanics and the neglect of spin by chance cancel each other in the case of the hydrogen atom.

Variational Principles and Wave Mechanics

THE great difficulties and serious shortcomings of the Bohr-Sommerfeld theory have been successfully eliminated in quantum mechanics, in which our whole attitude to such traditional concepts as space, time and matter underwent a radical revision. It is quite beyond the scope of this book to give even a cursory outline of wave or quantum mechanics, and we shall content ourselves with referring to those points that have a bearing on variational principles. A treatment of the formulation of quantum mechanics by means of an action principle will be given in the next section; in the present chapter we shall deal with a few miscellaneous topics.

In the original derivation of Schrödinger's wave equation he employed a variational principle, as well as the analogy of mechanics to optics mentioned earlier on. This derivation may therefore be briefly dealt with here. De Broglie[32] had already suggested that the Planck equation $E = h\nu$ might be of universal validity, so that with every material particle are associated waves of frequency determined by the above equation. On the basis of relativity mechanics, de Broglie postulated that with each particle is co-ordinated an "internal periodical phenomenon" of frequency $\nu_0 = \frac{1}{h} m_0 c^2$. To a fixed observer the frequency would therefore appear to be $\nu_0 \sqrt{1 - \frac{v^2}{c^2}}$. At the same time, he assumed that a wave of frequency $\frac{E}{h} = \frac{1}{h} \left(\frac{m_0 c^2}{\sqrt{1 - \frac{v^2}{c^2}}} \right)$ is also associated with the particle, and he showed that the two phenomena would always be in phase provided that the velocity of the wave is taken to be c^2/v. It follows that the wave-length is given by

$$\lambda = \frac{h}{\dfrac{m_0 v}{\sqrt{1 - \dfrac{v^2}{c^2}}}}$$

$$= \frac{h}{p}. \qquad . \qquad . \qquad . \qquad . \qquad (201)$$

This formula defines the "de Broglie wave-length." A most significant theorem issues from these considerations. For, calculating the group velocity u of these waves, we find

$$u = \frac{dv}{d\left(\dfrac{1}{\lambda}\right)}$$

$$= \frac{d\,\dfrac{1}{h}\left(\dfrac{m_0 c^2}{\sqrt{1 - \dfrac{v^2}{c^2}}}\right)}{d\,\dfrac{1}{h}\left(\dfrac{m_0 v}{\sqrt{1 - \dfrac{v^2}{c^2}}}\right)}$$

$$= v. \qquad . \qquad . \qquad . \qquad . \qquad (202)$$

According to this relation, the group of waves has the same velocity as the particle.

Supposing that the Planck equation and eq. (201) were satisfied, even in a potential field, de Broglie then verified that the paths of the particles are identical with the rays of the waves. To prove this, Fermat's principle may be written

$$\delta \int \frac{ds}{v\lambda} = 0, \qquad . \qquad . \qquad . \qquad (203)$$

or, since v is to be regarded as constant during the variation—a necessary restriction if the velocity of the waves depends upon the frequency—

$$\delta \int \frac{ds}{\lambda} = 0,$$

which results in

$$\delta \int \frac{m_0 v}{\sqrt{1 - \dfrac{v^2}{c^2}}}\, ds = 0. \qquad . \qquad . \qquad (204)$$

On the otherhand, the principle of least action becomes

$$\delta \int p \, ds = 0,$$

i.e.,
$$\delta \int \frac{m_0 v}{\sqrt{1 - \dfrac{v^2}{c^2}}} \, ds = 0.$$

Hence the two principles predict identical paths.

De Broglie now conceived of a promising interpretation of the Sommerfeld-Wilson conditions, a suggestion which originated from an entirely new standpoint. We recollect that the old theory had as its condition

$$\oint p \, dq = nh,$$

and, by substitution in accordance with de Broglie's hypothesis,

$$\oint \frac{h \, dq}{\lambda} = nh. \qquad . \qquad . \qquad . \qquad (205)$$

This equation simply implies that the path must contain a whole number of waves, or, that after one complete revolution, the phase of the waves must be unchanged.

These ideas of de Broglie commended themselves to Schrödinger, who developed them in a systematic and rigorous manner. De Broglie's explanation of the Bohr-Sommerfeld conditions could scarcely be assessed as anything more than a speculation, because it would necessitate the waves to possess no volume, but to exist only along the orbit. It was Schrödinger's object to find a suitable partial differential equation for the non-relativistic hydrogen atom.[19] Only those solutions were then to be allowed which were everywhere real, single-valued, finite and twice differentiable. The energies corresponding to these solutions would be the permissible energies. He commenced with the Hamilton-Jacobi equation

$$\left(\frac{\partial S}{\partial x}\right)^2 + \left(\frac{\partial S}{\partial y}\right)^2 + \left(\frac{\partial S}{\partial z}\right)^2 - 2m\left(E + \frac{e^2}{r}\right) = 0, \quad (206)$$

on which the conditions in the older quantum theory had been based. An equation of the type desired could not, however, be obtained from (206) as it stands, and Schrödinger consequently replaced S by $K \log \Psi$, so that the equation reads

$$\left(\frac{\partial \Psi}{\partial x}\right)^2 + \left(\frac{\partial \Psi}{\partial y}\right)^2 + \left(\frac{\partial \Psi}{\partial z}\right)^2 - \frac{2m}{K^2}\left(E + \frac{e^2}{r}\right)\Psi^2 = 0. \quad (207)$$

A differential equation of the required form could now be derived by the assumption that the variation of the integral of the left-hand side of (207), taken over all space, should vanish; on applying the Euler-Lagrange conditions, Schrödinger arrived at the equation

$$\nabla^2 \Psi + \frac{2m}{K^2}\left(E + \frac{e^2}{r}\right)\Psi = 0. \quad . \quad . \quad (208)$$

He determined the eigenvalues of this differential equation and, provided he took K to be $h/2\pi$, was led to the usual formula for the energy levels of the hydrogen atom, which give rise to the Balmer series.

In his second communication on wave mechanics,[19] Schrödinger devoted more attention to the derivation of (208). The transformation $S = K \log \Psi$ and the substitution of (207) by the variational principle did not seem to have any justification and were inserted only to produce a formally suitable differential equation. In order to establish a more secure foundation for his equation, he proceeded by methods related more closely to those of de Broglie, and introduced the Hamiltonian analogy between mechanics and optics, as outlined in §6. He recalled that geometrical optics had originally sufficed to account for the facts connected with the theory of light, but that, after further phenomena, e.g., interference and diffraction, had been discovered, ray optics was inadequate and an undulatory theory had to be adopted. It was conjectured by Schrödinger that a similar situation might prevail in mechanics, where the waves were supposed to be propagated in the $3n$-dimensional non-Euclidean space described in the exposition concerning his version of the relation between dynamics and optics. The classical view of a point travelling in this conceptual space is, he averred, only an approximation to a wave packet, comparable to the correspondence of rays and waves in optics. We immediately recognize here a fundamental difference between Schrödinger's conception and that of de Broglie, in whose theory the particle was *accompanied* and not *replaced* by waves.

As a requirement for the wave theory to be in harmony with classical dynamics in the limiting case of a wave packet, it is evident that the motion of such a packet must coincide with the motion of the corresponding point in configuration space. In

§6 we observed that the paths of the dynamical system will be identical with the rays of the wave system if the velocity of propagation of the waves, corresponding to an energy E, at any point is posited to be proportional to $\dfrac{1}{\sqrt{E-V}}$. The proportionality constant may, of course, involve the energy E. It is still necessary to verify that the group velocity of the waves is the same as the velocity of the dynamical system. This will be true as long as the phase velocity of the waves is taken to be $\dfrac{E}{\sqrt{2(E-V)}}$; the frequency, as usual, is to be specified by the Planck equation. The wave-length therefore becomes $\dfrac{h}{\sqrt{2(E-V)}}$. In the case of a single particle, this expression is equivalent to the de Broglie wave-length, as we are working in a co-ordinate system with the line element (80). Further, the directions of the waves and of the particle are identical, and the relation

$$p = \frac{h}{\lambda}$$

may be considered a vector equation. Returning to the general case, we have

$$u = \frac{d\nu}{d\left(\dfrac{1}{\lambda}\right)}$$

$$= \frac{d\,\dfrac{E}{h}}{d\left\{\dfrac{1}{h}\,\sqrt{2(E-V)}\right\}}$$

$$= \sqrt{2(E-V)}$$

$$= \sqrt{2T},$$

which is the velocity of the system, because we are assuming the line element (80).

It is interesting to trace how Schrödinger arrived at the above expression for the velocity. Beginning with the Hamilton-Jacobi equation (63), he pointed out that a solution is $W = S - Et$, where S is an integral of (62). The change of W with time may thus be treated as a motion of Schrödinger's surfaces of constant action, in such a manner that, after a time t, the surface S_0 has moved to the position initially occupied by the surface $S_0 + Et$. We have already demonstrated in §6 that the normal distance between the surfaces of constant action S_0 and $S_0 + dS$ is $\dfrac{dS}{\sqrt{2(E - V)}}$; it now follows automatically that the velocity of propagation of the W-waves is $\dfrac{E}{\sqrt{2(E - V)}}$. Schrödinger then surmised that the Ψ-waves travelled with the W-waves, and, in order for the frequency to be determined by the Planck equation, he set

$$\Psi = \sin\left(\frac{2\pi W}{h} + \text{const.}\right)$$

$$= \sin\left(-\frac{2\pi Et}{h} + \frac{2\pi S}{h} + \text{const.}\right).$$

This consideration seems irrelevant to the present investigation, but the same expression for the velocity could have been reached by a combination of the two conditions that the phase velocity must be proportional to $\dfrac{1}{\sqrt{E - V}}$, and that the group velocity has to agree with the velocity of the point representing the dynamical system.

We are now in a position to derive the renowned Schrödinger amplitude equation. The ordinary d'Alembert wave equation is

$$\nabla^2\Psi - \frac{1}{u_p{}^2}\frac{\partial^2\Psi}{\partial t^2} = 0,$$

where u_p denotes the phase velocity. Writing

$$\frac{\partial^2\Psi}{\partial t^2} = -4\pi^2\nu^2\Psi$$

$$= -4\pi^2\frac{E^2}{h^2}\Psi,$$

and $$u_p = \frac{E}{\sqrt{2(E - V)}},$$

we obtain

$$\nabla^2\Psi + \frac{8\pi^2}{h^2}(E - V)\Psi = 0. \qquad . \quad . \quad (209)$$

It must be borne in mind that we are still operating in non-Euclidean space with the line element given by (80). When we transform to our original generalized co-ordinates, (209) becomes

$$2T \left(q_r, \frac{\partial}{\partial q_r} \right) \Psi + \frac{8\pi^2}{h^2} (E - V) = 0,$$

or, since T is a homogeneous quadratic function of $\partial/\partial q_r$, we may write

$$H \left(q_r, \frac{h}{2\pi i} \frac{\partial}{\partial q_r} \right) \Psi = E\Psi, \qquad . \qquad . \quad (210)$$

which is the most frequently encountered form of the Schrödinger amplitude equation. In the case of the hydrogen atom, this equation reduces to (208).

In a later paper,[19] Schrödinger deduced a differential equation for the change of Ψ with respect to time. Eq. (210) holds for waves of frequency E/h, i.e., of waves involving the time through the factors $e^{2\pi i Et/h}$ or $e^{-2\pi i Et/h}$. Consequently,

$$\frac{\partial \Psi}{\partial t} = \pm \frac{2\pi i E}{h} \Psi,$$

and we arrive at the relation

$$H \left(q_r, \frac{h}{2\pi i} \frac{\partial}{\partial q_r} \right) \Psi = \pm \frac{h}{2\pi i} \frac{\partial \Psi}{\partial t}, \qquad . \qquad . \quad (211)$$

which is Schrödinger's original formulation of his wave equation. Eq. (210) is, according to Schrödinger's theory, the condition that the wave function has frequency E/h, or that the corresponding system possesses energy E.

From (211), a system represented by a wave packet, whose space dependence is expressed by $Ae^{2\pi i \mathbf{p.x}/h}$, will move in either direction with momentum $\mp \mathbf{p}$. It is thereby strongly suggested that we replace the $\mp$ sign in this equation by a $-$ sign, so that

$$H \left(q_r, \frac{h}{2\pi i} \frac{\partial}{\partial q_r} \right) \Psi = - \frac{h}{2\pi i} \frac{\partial \Psi}{\partial t}. \qquad . \qquad . \quad (212)$$

In this event, a wave packet whose space dependence is $Ae^{2\pi i \mathbf{p.x}/h}$ would travel with momentum $\mathbf{p}$, whereas a wave packet $Ae^{-2\pi i \mathbf{p.x}/h}$ would travel with momentum $-\mathbf{p}$. The Schrödinger wave equation is therefore essentially different from that of d'Alembert in that it is linear in the time, and Ψ is thus known for all values of t provided it is defined initially. Once the initial motion is specified, the path of a wave packet is the same whether we use (212) or (211), the sign merely determining the direction of motion, and the correspondence between wave and classical mechanics is still assured.

Having completed the review of the derivation of the Schrödinger wave equation and the rôle which the classical concept of action played in it, we can pass on to the use of this concept in establishing certain relations between quantum and classical mechanics. A function W is sometimes defined by the formula

$$\Psi = A e^{2\pi i W / h}, \qquad . \qquad . \qquad . \qquad (213)$$

where A and W are real numbers. The justification of this definition is that, for a wave packet, eq. (213) may be written

$$\Psi = A e^{2\pi i (\mathbf{x}.\text{grad } W \,+\, \text{const.})/h} \; e^{2\pi i \frac{\partial W}{\partial t} \, t/h} . \qquad . \qquad (214)$$

The derivatives of W are supposed to vary slowly over distances comparable to the wave-length, and the above relation will consequently be fulfilled very closely over a large number of waves. Eq. (214) together with the de Broglie and Planck formulae result in the equations

$$\frac{\partial W}{\partial x_r} = p_r ; \qquad \frac{\partial W}{\partial t} = -E.$$

It follows that W is the Hamilton principal function for the classical particle corresponding to the wave packet under consideration. If we apply the Schrödinger wave equation to (213) and neglect all terms which contain h, we are left with the Hamilton-Jacobi equation. All the imaginary terms in the Schrödinger equation applied to (213) involve h, and, by equating the terms linear in h, we obtain an equation for A which can be used to show that the motion of the wave packet is the same as that of the classical particle which it represents. A function W', which is also occasionally employed, is defined by the equation

$$\Psi = e^{2\pi i W'/h},$$

whence we deduce that

$$W' = W + \frac{h}{2\pi i} \log A$$

and that W', which differs from W by a quantity of order h, still approximates, for a wave packet, to the Hamilton principal function.

In a like manner, we may define a function S from a solution Ψ of the Schrödinger amplitude equation by the formula

$$\Psi = A e^{2\pi i S/h}, \qquad . \qquad . \qquad . \qquad (215)$$

and, if h is taken to be small, it can be verified as before that the

Hamilton-Jacobi equation (62) is satisfied. This action function S is not of much avail for the theory of wave packets, as such a packet is of necessity a combination of different energy states; it can be used, however, to indicate the relation between the Sommerfeld-Wilson quantum conditions and Schrödinger's amplitude equation. We first of all recognize that, when S can be expressed as the sum of functions, each involving one of the co-ordinates only, Ψ may be written as the product of such functions. The stipulation of single-valuedness entails that, after a complete oscillation of one of the co-ordinates, the value of Ψ must be the same as it was at the beginning. Since S continually increases, it cannot repeat its value and must therefore change by an integral multiple of h. It is to be understood that an argument of this type can in no sense be regarded as a strict mathematical proof, because it has not been confirmed that the function S of (215) is the Hamilton characteristic function for some path of the corresponding classical system. Besides, in deriving the Hamilton-Jacobi equation from (215), we have assumed h to be small, a presupposition we have no right to make for ordinary quantum states. The method of reasoning, nevertheless, suggests a connexion between the quantization of phase integrals and the Schrödinger amplitude equation. A more definite demonstration would be impossible: the older theory does not always lead to the same results as quantum mechanics, e.g., for the problem of the relativistic hydrogen atom without spin.

Before concluding this section, we must call attention to the point that Schrödinger's amplitude equation is an instance of the "Sturm-Liouville" type which can be expressed as a variational principle. For, if we apply the Euler-Lagrange equations to the expression

$$\frac{h^2}{4\pi^2} T\left(q_r, \frac{\partial \Psi}{\partial q_r}\right) + V(q_r)\Psi^2 - E\Psi^2,$$

we get eq. (210), as T is a homogeneous quadratic function of the derivatives. Indeed, it was from such reflections that the Schrödinger equation was originally derived. Again, it is known from the elementary theory of the Sturm-Liouville equation that (210) is the condition for the integral

$$\int \left\{ \frac{h^2}{4\pi^2} T\left(q_r, \frac{\partial \Psi}{\partial q_r}\right) + V(q_r)\Psi^2 \right\} d\tau$$

to be stationary, subject to the restriction that Ψ is normalized. Likewise, if (210) holds, the above integral is a minimum, subject to the restrictions that Ψ is normalized and is in addition orthogonal to all the eigenfunctions of this equation which have energy less than E. These theorems are sometimes of assistance in approximating to the eigenvalues of the Schrödinger equation. It must not be overlooked, however, that any second-order linear differential equation can be reduced to the Sturm-Liouville type and thence formulated as a variational principle; we should therefore exercise reserve in attributing more than pragmatic significance to such variational interpretations of wave-mechanical equations.

Postscript

We are indebted to Professor E. Schrödinger for having prompted the following comment.

The quantum mechanical Hamiltonian appearing in (210) conforms to the prescription $p_r \rightarrow \dfrac{h}{2\pi i}\dfrac{\partial}{\partial q_r}$ found in most treatises on quantum mechanics. It must be remembered, however, that if the co-ordinates are not Cartesian, or if the forces are momentum-dependent, this expression is ambiguous, owing to the non-commutability of q_r and $\dfrac{\partial}{\partial q_r}$.

For a general Riemannian space with a metric

$$ds^2 = \sum_{r,s} g_{rs} dq_r dq_s,$$

the Laplacian becomes, according to a well-known result from tensor calculus,

$$\nabla^2 = \sum_{r,s} g^{-\frac{1}{2}} \frac{\partial}{\partial q_r}\left(g^{\frac{1}{2}}\, g^{rs}\, \frac{\partial}{\partial q_s}\right),$$

where g is the determinant of the matrix g_{rs} and the g^{rs} stand for the elements of the reciprocal matrix.

This equation holds in particular for the Laplacian in eq. (209), so that in general the Hamiltonian which is to be substituted in eq. (210) has the unique form

$$H = -\tfrac{1}{2}\left(\frac{h}{2\pi}\right)^2 \sum_{r,s} g^{-\frac{1}{2}}\frac{\partial}{\partial q_r}\left(g^{\frac{1}{2}}\, g^{rs}\, \frac{\partial}{\partial q_s}\right) + V.$$

To find an explicit expression for g^{rs} it is only necessary to equate the above expression for ds^2 with the value of the metric given in eq. (80); we obtain, after dividing by dt^2,

$$\sum_{r,s} g_{rs}\dot{q}_r\dot{q}_s = 2\bar{T}(q_r,\dot{q}_r).$$

If we substitute on the right-hand side for T its definition and use eq. (34), comparison of the coefficients of $\dot{q}_r\dot{q}_s$ leads to the required results:

$$g_{rs} = \sum_i \sum_{x,y,z} m_i \frac{\partial x_i}{\partial q_r} \frac{\partial x_i}{\partial q_s},$$

$$g^{rs} = \sum \sum_{x,y,z} \frac{1}{m_i} \frac{\partial q_r}{\partial x_i} \frac{\partial q_s}{\partial x_i}.$$

§ 12

The Principles of Feynman and Schwinger in Quantum Mechanics

THE foregoing survey has borne out the major part which the action function has played in the borderline problems between classical and wave mechanics. We now propose to examine recent formulations of the laws of quantum mechanics themselves by means of action principles. Such formulations have been evolved by Feynman[33] and Schwinger,[34] and they are logically preferable to the now traditional canonical equations for several reasons. Strictly speaking, neither of these original theorems is a "stationary" principle of the type described in the foregoing chapters; in fact, Feynman's expression for the transition amplitude between eigenstates does not represent a variational principle at all. However, within the limits in which classical mechanics prevails, these new formulations reduce to the principle of least action—indeed, they may even be regarded as its proper quantum analogue. It will therefore be appropriate to give in this section a brief exposition of the Feynman and Schwinger approaches.

The ordinary Hamiltonian presentation of quantum mechanics contains, apart from the basic principles of the subject, two distinct postulates: the commutation relations between generalized co-ordinates and momenta, and the equations of motion. We have already noted that, in classical mechanics, the Lagrangian formalism, which does not involve the generalized momenta, can be substituted for the Hamilton theory. The laws of mechanics may then be expressed by a single postulate, which can alternatively assume the form of Lagrange's equations or of the principle of least action as stated by Hamilton. The obvious question that arises from such considerations: is it not possible, and moreover most desirable, to obtain a similar Lagrangian formulation of the quantum laws depending upon a single postulate only? This aim is achieved by the Feynman and Schwinger theories. One cannot fail to observe that Feynman's principle in particular—and this is no hyperbole—expresses the laws of quantum mechanics in an exemplarily neat and elegant manner, notwithstanding the fact that it employs somewhat unconventional mathematics. It can easily be related to Schwinger's principle, which utilizes mathematics of a more familiar nature. The theorem of Schwinger is, as it were, simply a translation of that of Feynman into differential notation.

In quantum field theory, these novel formulations possess a definite, additional advantage over those which preceded them. The Hamiltonian version of quantum field theory is not cast in a manifestly co-variant form, and it has to be verified in each case separately that the theory does satisfy the invariance requirements of special relativity. This seldom presents great difficulty, but it would be advantageous to construct a theory in which it is immediately conspicuous that the postulates of relativity are satisfied. In the Feynman and Schwinger schemes, the relativistic co-variance of the theory depends upon the relativistic invariance of the Lagrangian alone.

We shall first concern ourselves with the Feynman principle and continue by showing how it is related to the canonical equations of quantum mechanics. The connexion of this principle with the law of least action in classical mechanics will also be indicated. Finally, we shall describe the manner in which Schwinger's results derive from those of Feynman.

A dynamical system in quantum mechanics is defined uniquely by means of a complete set of commuting observables, such as the generalized co-ordinates, q_r. One can then adopt the simultaneous eigenstates of all the q_r's as the basic states, in terms of which any arbitrary state of the system can be expressed. We suppose that all the operators q_r's have continuous eigenvalues and denote a simultaneous eigenstate, in which q_r has the eigenvalue q_r, by $|q_r\rangle$. Any state of the system, written as an integral of eigenstates of q_r's, thus becomes

$$\int \Psi(q_r)|q_r\rangle dq_r,$$

where Ψ is a complex function of the q_r's. In the special case when the q_r's are the space co-ordinates of a particle, Ψ is nothing but the ordinary Schrödinger wave function.

The q_r's will change with time, and if q_{r1} and q_{r2} denote q_r at times t_1 and t_2 respectively, the state $|q_{r1})$ will be different from the state $|q_{r2})$. The transition amplitude between the states $|q_{r1})$ and $|q_{r2})$ is a complex number which will be symbolized by $(q_{r2}|q_{r1})$, so that $(q_{r2}|q_{r1})$ is defined by

$$|q_{r1}) = \int |q_{r2})(q_{r2}|q_{r1})dq_{r2}. \qquad \cdot \qquad \cdot \qquad (216)$$

As a consequence, a knowledge of $(q_{r2}|q_{r1})$ suffices to determine the dynamical behaviour of the system with time. The wave function $\Psi(q_{r2})$ can be obtained from $\Psi(q_{r1})$ using the formula

$$\Psi(q_{r2}) = \int (q_{r2}|q_{r1})\Psi(q_{r1})dq_{r1}, \qquad \cdot \qquad \cdot \qquad (217)$$

which follows easily from (216) and the definition of Ψ.

A problem which now suggests itself is the determination of the transition amplitudes between the times t_1 and t_3, once it is known between t_1 and an intermediate time t_2, and between t_2 and t_3. This question is answered forthwith by the well-known quantum-mechanical equation

$$(q_{r3}|q_{r1}) = \int (q_{r3}|q_{r2})(q_{r2}|q_{r1})dq_{r2}. \qquad \cdot \qquad \cdot \qquad (218)$$

We could view $(q_{r3}|q_{r2})(q_{r2}|q_{r1})$ as the transition amplitude from the state $|q_{r1})$ at time t_1 to the state $|q_{r3})$ at the time t_3 via the state $|q_{r2})$ at time t_2; the total transition amplitude between the state $|q_{r1})$ at time t_1 and the state $|q_{r3})$ at time t_3 is then found by integrating the above amplitude over all the intervening states $|q_{r2})$. Instead of dividing the interval between the initial and final times only once, we may subdivide it a number of times; a repeated application of eq. (218) will accordingly produce the formula

$$(q_{rn}|q_{r1}) = \int (q_{rn}|q_{r(n-1)})(q_{r(n-1)}|q_{r(n-2)}) \cdots$$
$$(q_{r2}|q_{r1})dq_{r2}dq_{r3} \cdots dq_{r(n-1)}. \qquad \cdot \qquad (219)$$

This equation thus decomposes the transition amplitude between the states $|q_{r1})$ at the instant t_1 and $|q_{rn})$ at the instant t_n into amplitudes governing the passage of the system along the path in configuration space for which q_r has the value q_{r1} at time t_1, q_{r2} at time t_2, . . ., q_{rn} at time t_n. The total transition amplitude is then obtained by integration of the partial amplitudes over q_{r2}, . . ., $q_{r(n-1)}$.

The values of q_r at the times t_1, t_2, . . ., t_n still only specify the path at a discrete and finite number of instants. Feynman now conceives this subdivision to be carried out indefinitely, so that the total transition amplitude becomes a sum of elementary contributions, one from each continuous trajectory passing through

q_{r1} at time t_1 and q_{rn} at time t_n. Next, he proposes that the equations of motion of the system be expressed by· stating the values of these elementary transition amplitudes. He formulates his principle as follows: *The transition amplitude between the states* $|q_{r1}\rangle$ *and* $|q_{rn}\rangle$ *of a quantum-mechanical system is the sum of elementary contributions, one from each trajectory passing between* q_{r1} *at time* t_1 *and* q_{rn} *at time* t_n. *Each of these contributions has the same modulus, but its phase is the classical action integral* $\int L dt$ *(in units of* $\hbar$*)*[†] *for the path.* In symbols,

$$(q_{rn}|q_{r1}) = \frac{1}{N} \int e^{i\int_{t_1}^{t_n} L dt} \, \delta q_r(t). \qquad . \qquad . \quad (220)$$

The differential $\delta q_r(t)$ indicates that we must integrate over all paths connecting the given initial- and end-points q_{r1} at $t = t_1$ and q_{rn} at $t = t_n$ respectively. The factor $1/N$ is simply a normalizing factor which will be determined later.

It may be worth while to remark *en passant* upon the mathematical procedure implied by eq. (220). The novel aspect resides in the fact that instead of integrating over a finite number of variables, we integrate over *all paths*, and a path requires an infinite number of variables for its complete specification—e.g., the values of the q_r's at all the values of t between t_1 and t_n. The exponent $\int L dt$ in the integrand is the action integral for a particular path, i.e., for a specific set of values which the infinity of variables $q_r(t)$ can assume. It depends thus upon each of the infinite aggregate of integration variables. Consequently, it becomes necessary to define the integral in (220) as the limit of an integration over a finite number of variables. Feynman himself treats of this problem by subdividing the interval t_1, t_n into a large number of intervals of magnitude $t_2 - t_1$, $t_3 - t_2$, . . ., $t_n - t_{n-1}$. The path is then specified by the values of the variables q_{r2}, q_{r3}, · · ·, $q_{r(n-1)}$, of which there are a large but finite number. The action integral $\int L(q_r, \dot{q}_r) dt$ in the integrand is accordingly replaced by the sum

$$\sum_{i=1}^{n-1} \mathcal{L}(q_{r(i+1)}, q_{ri})(t_{i+1} - t_i)$$
$$\equiv \sum_{i=1}^{n-1} L\left(\frac{q_{r(i+1)} + q_{ri}}{2}, \frac{q_{r(i+1)} - q_{ri}}{t_{i+1} - t_i}\right)(t_{i+1} - t_i),$$

[†] In this section we make a slight change of terminology and refer to Hamilton's integral $\int L dt$ as the "action." We shall also designate Hamilton's principle as "the principle of least action."

where the generalized co-ordinates and velocities in the Lagrangian for each interval are replaced by their mean values over the interval. If the number of subdivisions is large, the difference between this sum and the true action integral over the trajectory is small. In agreement with this fact, the integral in (220) is approximated by the integral

$$\int e^{i\sum_{i=1}^{n-1}\mathcal{L}(q_{r(i+1)}, q_{ri})(t_{i+1} - t_i)} \frac{1}{A_1} \frac{dq_{r2}}{A_2} \cdots \frac{dq_{r(n-1)}}{A_{n-1}}. \tag{221}$$

The factors $1/A$ correspond to the normalization factor $1/N$; they will be found to depend on the size of the time intervals and must therefore be inserted at this stage. Finally, the limit of (221), as the number of subdivisions tends to infinity, is supposed to represent the required transition amplitude.

At first sight, it may appear that completely discontinuous paths contribute to the same extent as continuous paths in (221), since there is no factor restricting $q_{r(i+1)}$ to be near q_{ri}. However, L will usually involve powers of $\dot{q}_r$ greater than the first, and the complex exponential in eq. (221) will thus be large and rapidly varying for paths in which $q_{r(i+1)}$ is not near to q_{ri}. It can be seen that these paths cancel one another out in the integral, and only continuous paths remain to contribute appreciably.

It is not merely through the subdivision process outlined above that the integral (220) can acquire significance. We may, in principle, use any method of specifying the path, provided that a valid limiting process exists. The functions $q_r(t)$ can, for instance, be expanded into a complete set of functions obeying the given boundary conditions on the co-ordinates. The expansion coefficients are then to be regarded as the new variables of integration. By discarding all but the leading n components of this resolution, the number of integration variables can be rendered finite; in the final stage of the process, n is allowed to tend to infinity. Again, as mentioned earlier, the introduction of the normalization factor should precede the transition to the limit of an infinite number of integration variables.

Although extensive work has been done on the theory of functionals, i.e., of functions of an infinite number of variables, not much attention has up to the present been devoted to functional integration. The convergence of the limiting process just discussed has not been investigated, nor has it been proved that the result is independent of the variables employed to define the path. (It is, of course, understood that one does not transform to a new set of variables without simultaneously multiplying the integrand by the appropriate Jacobian.) We encounter here a decidedly unfamiliar mathematical topic which will no doubt become the subject of extensive research, once the Feynman formalism plays a prominent role in the development of physical theory. Indeed, as soon as we have surmounted the difficulties peculiar to the new mathematical concepts involved, eq. (220) provides us with an extremely succinct presentation of the laws of quantum mechanics.

We shall now demonstrate the equivalence between the Feynman formulation of quantum mechanics and the canonical formalism, but it is necessary first to establish two preliminary results: (a) the value of the normalization factor $1/A$ in (221), (b) the formula for the matrix element of an operator in the Feynman method.

The normalization factor can be derived from the equation for the change of Ψ with time. From (217), using (220) and (221),

$$\Psi(q_{rn}) = \int \Psi(q_{r1}) e^{i \sum\limits_{i=1}^{n-1} \mathcal{L}(q_{r(i+1)}, q_{ri})(t_{i+1} - t_i)} \frac{dq_{r1}}{A_1} \frac{dq_{r2}}{A_2} \cdots \frac{dq_{r(n-1)}}{A_{n-1}}. \quad (222)$$

Likewise,

$$\Psi(q_{r(n+1)}) = \int \Psi(q_{r1}) e^{i \sum\limits_{i=1}^{n} \mathcal{L}(q_{r(i+1)}, q_{ri})(t_{i+1} - t_i)} \frac{dq_{r1}}{A_1} \frac{dq_{r2}}{A_2} \cdots \frac{dq_{rn}}{A_n}, \quad (223)$$

and therefore

$$\Psi(q_{r(n+1)}) = \int \Psi(q_{rn}) e^{i\ (q_{r(n+1)}, q_{rn})(t_{n+1} - t_n)} \frac{dq_{rn}}{A_n}. \quad (224)$$

We further note that the rate at which the exponential oscillates, increases as the interval $t_{n+1} - t_n$ decreases in size; and since the wave function Ψ varies very little by comparison, it may be placed in front of the integral sign. Hence

$$\Psi(q_{r(n+1)}) = \Psi(q_{rn}) \int e^{i\mathcal{L}(q_{r(n+1)}, q_{rn})(t_{n+1} - t_n)} \frac{dq_{rn}}{A_n}.$$

When t_{n+1} tends to t_n, $\Psi(q_{r(n+1)})$ tends to $\Psi(q_{rn})$, so that we obtain finally

$$A_n = \int e^{i\,\mathcal{L}(q_{r(n+1)}, q_{rn})(t_{n+1} - t_n)}\, dq_{rn}. \quad (225)$$

In the special case where L has the form

$$L = \sum_r \tfrac{1}{2} a_r(q_s) \dot{q}_r{}^2 - V(q_s), \quad (226)$$

eq. (225) becomes

$$A_n = \prod_r \{2\pi(t_{n+1} - t_n)/a_r(q_s)\}^{\frac{1}{2}}. \qquad . \qquad . \qquad (227)$$

A thus contains a weight factor which in general differs from point to point in configuration space. (If, for instance, polar co-ordinates are used, the weight factor of A simply multiplies the differential $d\theta d\phi$ by r^2.) In addition, there appear in A factors proportional to the square root of the difference $t_{n+1} - t_n$, one for each degree of freedom, if the homogeneous quadratic form mentioned above is assumed for the Lagrangian.

We find the matrix element of an operator in the Feynman formalism by generalizing the fundamental equation (220). What we require is an expression of the type

$$(q_{rn}|O(q_{rm})|q_{r1}),$$

where, as usual, $|q_{r1})$ and $|q_{rn})$ denote simultaneous eigenstates of the q_r's at the times t_1 and t_n respectively and the argument q_{rm} of the operator O signifies that the co-ordinates have to be taken at an intermediate time t_m. By expanding this expression according to (216),

$$(q_{rn}|O(q_{rm})|q_{r1}) = \int (q_{rn}|O(q_{rm})|q_{rm})(q_{rm}|q_{r1})dq_{rm}$$

$$= \int O(q_{rm})(q_{rn}|q_{rm})(q_{rm}|q_{r1})dq_{rm}. \qquad . \qquad (228)$$

The transition amplitudes in (228) can now be written as functional integrals by substitution from Feynman's formula (220), so that

$$(q_{rn}|O(q_{rm})|q_{r1}) = \frac{1}{N_1}\frac{1}{N_2} \int O(q_{rm}) e^{i\int_{t_1}^{t_m} Ldt} e^{i\int_{t_m}^{t_n} Ldt} \delta_{t_1 < t < t_m} q_r(t)\delta_{t_m < t < t_n} q_r(t)dq_{rm}. \qquad . \qquad (229)$$

The notation $\delta_{t_1 < t < t_m}$ indicates that we must integrate over all paths with end-points q_{r1} at t_1 and q_{rm} at t_m respectively, whereas $\delta_{t_m < t < t_n}$ implies that the integration has to be extended over all paths which terminate at q_{rm} at t_m and q_{rn} at t_n respectively. In other words, we are allowed to include only trajectories which intersect in q_{rm} at the instant t_m. This last restriction is, however, removed by the subsequent integration over q_{rm}, and eq. (229) may, consequently, be restated as

$$(q_{rn}|O(q_{rm})|q_{r1}) = \frac{1}{N} \int O\{q_r(t_m)\} e^{i\int_{t_1}^{t_n} Ldt} \delta q_r(t). \qquad . \qquad (230)$$

To compute the matrix element of the operator $O(q_{rm})$ connecting the states $|q_{r1})$ and $|q_{rn})$, we take a contribution $e^{i \int L dt}$ from each path and multiply it by the value of $O(q_{rm})$ along the path under consideration. By an application of (219), eq. (230) can be generalized to cover the case where the operator is a product of separate operators pertaining to different times, provided that the factors are arranged in chronological order, reading from right to left. Our equation will then assume the form

$$(q_{rn}|T\{O(q_{rk})O'(q_{rl}) \ . \ . \ . \ O''(q_{rm})\}|q_{r1})$$

$$= \frac{1}{N} \int O\{q_r(t_k)\}O'\{q_r(t_l)\} \ . \ . \ . \ O''\{q_r(t_m)\}e^{i\int_{t_1}^{n} L dt} \, \delta q_r(t), \qquad (231)$$

where the symbol T outside the curly bracket ensures the chronological order of the operators following it. Equations (230) and (231) are generalizations (or rather extensions) of the formula (220), but they need not be included in the fundamental postulate which (220) symbolizes, since they can be deduced from it in virtue of the arguments expounded above.

The proof that the Feynman approach is equivalent to the traditional presentation of quantum mechanics is now a relatively easy matter, because both the commutation rules and equations of motion in canonical formalism flow directly from the equations derived in the preceding paragraphs. In the commutation relations the momenta will not appear as such, but only as derivatives of the Lagrangian with respect to the velocities ($P_r = \partial L/\partial \dot{q}_r$, in analogy to the classical equation (46)). The equations of motion will be obtained in the usual Lagrangian form.

We proceed by considering the matrix element of the derivative of an operator with respect to one of the co-ordinates; thus,[†]

$$\left(q_{rn}\left|\frac{\partial O_m}{\partial q_{rm}}\right|q_{r1}\right) = \frac{1}{N}\int \frac{\partial O_m}{\partial q_{rm}} e^{i\int_{t_1}^{t_n} L dt} \, \delta q_r(t)$$

$$= \frac{1}{N}\int \frac{\partial}{\partial q_{rm}}\left\{O_m e^{i\int_{t_1}^{n} L dt}\right\} \delta q_r(t) - \frac{i}{N}\int O_m \left(\frac{\partial}{\partial q_{rm}}\int_{t_1}^{t_n} L dt\right) e^{i\int_{t_1}^{n} L dt} \, \delta q_r(t).$$

† For simplicity, we assume that N is independent of the q_r's. The result (233) can also be derived if N is a function of the q_r's, but a measure of caution is to be exercised.

On integration over all paths, the first term vanishes, and, transforming the second term according to Feynman's formula (230), we are left with

$$\left(q_{rn} \left| \frac{\partial \boldsymbol{O}_m}{\partial \boldsymbol{q}_{rm}} \right| q_{r1} \right) = - i \left(q_{rn} \left| T \left\{ \boldsymbol{O}_m \left(\frac{\partial}{\partial \boldsymbol{q}_{rm}} \int_{t_1}^{t_n} L dt \right) \right\} \right| q_{r1} \right). \quad (232)$$

Since the eigenstates $|q_{r1})$ and $|q_{rn})$ are arbitrary and belong to a complete set of states, eq. (232) remains valid also as an operator equation, i.e.,

$$\frac{\partial \boldsymbol{O}_m}{\partial \boldsymbol{q}_{rm}} = - i T \left\{ \boldsymbol{O}_m \left(\frac{\partial}{\partial \boldsymbol{q}_{rm}} \int_{t_1}^{t_n} L dt \right) \right\}. \quad . \quad . \quad (233)$$

This equality holds generally for the derivatives of an operator; as a special case we may take $\boldsymbol{O}$ to stand for one of the co-ordinates $\boldsymbol{q}_s$. The left-hand side is then simply δ_{rs}. We can therefore deduce the commutation relations by proving the identity of the right-hand side with the commutator of $\boldsymbol{q}_{sm}$ and $\partial L/\partial \dot{\boldsymbol{q}}_{rm}$. The expression on the right of (233) depends only upon the contributions to the action integral from times in the immediate vicinity of t_m, so that, if the integral is approximated by a sum, the equation becomes

$$q_{sm} \frac{\partial}{\partial \boldsymbol{q}_{rm}} L \left(\frac{\boldsymbol{q}_{rm} + \boldsymbol{q}_{r(m-1)}}{2}, \frac{\boldsymbol{q}_{rm} - \boldsymbol{q}_{r(m-1)}}{t_m - t_{m-1}} \right) (t_m - t_{m-1})$$

$$+ \left\{ \frac{\partial}{\partial \boldsymbol{q}_{rm}} L \left(\frac{\boldsymbol{q}_{r(m+1)} + \boldsymbol{q}_{rm}}{2}, \frac{\boldsymbol{q}_{r(m+1)} - \boldsymbol{q}_{rm}}{t_{m+1} - t_m} \right) \right\} (t_{m+1} - t_m) \boldsymbol{q}_{sm}$$

$$= i \delta_{rs}. \quad . \quad . \quad . \quad (234)$$

(We note that the operators have been placed in the correct chronological order.) On neglecting the terms proportional to the differences $t_m - t_{m-1}$ and $t_{m+1} - t_m$, eq. (234) immediately simplifies to the commutation relations

$$q_{sm} \frac{\partial L}{\partial \dot{\boldsymbol{q}}_{rm}} - \frac{\partial L}{\partial \dot{\boldsymbol{q}}_{rm}} \boldsymbol{q}_{sm} = i \delta_{rs}. \quad . \quad . \quad (235)$$

To take another special case, the equations of motion follow from (233) if $\boldsymbol{O}$ is put equal to a constant. The time-ordering operator can now be ignored, and we are left with the derivative of the action integral with respect to the rth co-ordinate at time t_m. The reader will recall that we have already considered such a functional derivative in the steps leading to eq. (112); if we avail ourselves of the same arguments here, we can show that

$$\frac{\partial}{\partial \boldsymbol{q}_{rm}} \int_{t_1}^{t_n} L dt \simeq \left(\frac{\partial \boldsymbol{L}_m}{\partial \boldsymbol{q}_{rm}} - \frac{d}{dt} \frac{\partial \boldsymbol{L}_m}{\partial \dot{\boldsymbol{q}}_{rm}} \right) (t_{m+1} - t_m),$$

provided we assume that the time differences $t_{m+1} - t_m$ and $t_m - t_{m-1}$ are equal. On substitution, eq. (233) thus reads

$$\frac{\partial \boldsymbol{L}}{\partial \boldsymbol{q}_{rm}} - \frac{d}{dt} \frac{\partial \boldsymbol{L}}{\partial \dot{\boldsymbol{q}}_{rm}} = 0,$$

which are nothing but Lagrange's equations of motion in operator language. Hence both the commutation rules and the equations of motion are logical consequences of the Feynman formalism, the equivalence of which with the canonical formalism is thereby completely established.

With a modicum of reasoning, an analogy between Feynman's basic equation (220) and Huygens' optical principle suggests itself. In fact, the following simple physical interpretation of (220) would be warranted. Imagine a source at the point q_{r1} to be emitting waves of all velocities in such a manner that, after a short time interval, the amplitudes of all the waves are equal, but each is retarded by a phase-difference equal to the action taken over the distance traversed. Each point on the new wave fronts will then become a secondary source of waves during the next time interval. The cumulative effect of like processes is that waves can travel from q_{r1} at time t_1 to q_{rn} at time t_n along any path, and the contribution from each path is retarded in phase by an amount equal to the action integral calculated for that particular path. The total amplitude at q_{rn} is obtained by adding the individual contributions from all the paths. The optical analogue therefore leads cogently to the Feynman postulate (220).

The change of a dynamical system with time in quantum mechanics can accordingly be formulated concisely and comprehensively by means of Huygens' principle; it is perhaps worth mentioning that this is the only discipline of physics which is susceptible to a consistent treatment by Huygens' concept. The wave equations of classical physics are differential equations of the second order with respect to time, whereas Huygens' principle—it determines the wave function at any time once it is known throughout space at an earlier time—is applicable without modification exclusively to first-order differential equations. The Feynman principle differs from Huygens' theory as originally conceived in that the action replaces the time in determining the phase retardation. This was to be expected, since the wave surfaces, as we have seen before, are the surfaces of constant action for the corresponding corpuscular system.

To avoid the risk of over-simplification, we should stress that the resemblance between the Feynman and Huygens principles is, in a sense, mathematical rather than physical, because the space in which the wave propagation is supposed to occur is not physical space, but the n-dimensional configuration space of the quantum-mechanical system.

Finally, it remains to be shown that, when the limit of classical mechanics is approached, Feynman's principle transforms gradually into the principle of least action. In this limit, the probability amplitude will differ appreciably from zero only for those values of the generalized co-ordinates at a given time t_1, which are close to the corresponding values for the classical system; let them be denoted by q_{r1}. At a later time t_n, the wave packet will have travelled to q_{rn}. The Feynman formula expresses the transition amplitude $(q_{rn}|q_{r1})$ as the sum of transition amplitudes for all possible paths connecting q_{r1} and q_{rn}. Consider now that in the limit envisaged here, each path will contain a large number of units of action, so that the phase will vary rapidly as we change from one path to another. The contributions made to the integral by different paths will therefore cancel one another out, and the only significant remaining contribution derives from trajectories in that region of configuration space where the action is stationary for variations of the path—in other words, in the region surrounding the classical path. We have therefore reduced the total transition amplitude virtually to the sum of partial amplitudes arising from the classical and adjacent paths: that is, in the classical limit the wave packet moves in accordance with the principle of least action.

The foregoing discussion shows how Feynman's principle, although it is actually not a variational principle, does none the less generate, in a direct and consummate fashion, a variational principle, when the conditions are such that classical mechanics provides a valid description of the system under consideration. The position is precisely analogous to that in optics, where Fermat's principle—despite its failure to embrace optics as a whole—suffices to furnish the laws of geometrical optics.

The Feynman theory of quantum mechanics distinguishes itself from all other concurrent formulations in that it rests upon an integral instead of a differential principle: it associates a probability amplitude $(q_{rn}|q_{r1})$ with the entire development of a quantum mechanical system between two configurations, rather than with the configuration of the system at a particular instant. This has been achieved at the expense of sanctioned mathematical procedure. Schwinger's approach involves, on the other hand, a differential treatment, and it lacks perhaps some of the elegance of Feynman's method. Still, it does not require the use of an *ad hoc* mathematical technique. Further, Schwinger's theory retains the paramount advantage of Feynman's formulation: it stems from a single postulate, in contradistinction to the two postulates of the canonical method, and it expresses the basic laws of quantum mechanics in terms of a single function—the action.

Although Schwinger evolved his theory independently, we shall present it here as a consequence of the Feynman principle; we have already remarked at the beginning of this section that the present theory is merely a translation into differential form of the preceding one. Schwinger inquires how the transition amplitude is modified when infinitesimal changes occur in the dynamical variables of the quantum-mechanical system. These changes may be of various kinds. We shall, however, restrict ourselves to the case where they are of the nature of those treated earlier, in the section on classical mechanics, viz., variations in the co-ordinates q_r, and variations in the time which sends the operator $q_r(t)$ into $q_r(t + \delta t)$. In addition, we shall be satisfied with c-number variations of the co-ordinates. The change in the transition amplitude between the states $|q_{r1})$ and $|q_{rn})$ follows immediately on varying Feynman's fundamental relation (220):

$$\delta(q_{rn}|q_{r1}) = \frac{i}{N} \int \delta \left(\int_{t_1}^{t_n} L\,dt \right) e^{i\int_{t_1}^{t_n} L\,dt} \, \delta q_r(t). \qquad . \qquad (236)$$

Eq. (230) allows us to write the right-hand side as a matrix element, so that (236) becomes

$$\delta(q_{rn}|q_{r1}) = i \left(q_{rn} | \delta \int_{t_1}^{t_n} L\,dt | q_{r1} \right). \qquad . \qquad . \qquad (237)$$

We have here, in symbolical form, Schwinger's variational principle which enunciates: *if variations are effected in a quantum-mechanical system, the corresponding change in the transition amplitude between the eigenstates $|q_{r1})$ and $|q_{rn})$ is i times the matrix element connecting the two states of the variation of the action integral* $\int L\,dt$. This formulation of the principle is incomplete unless we qualify more fully the type of variation for which it is valid. It will be shown below that it is sufficient for the purpose of deriving the laws of quantum mechanics to consider only c-number variations of co-ordinates and time. Schwinger's principle holds, however, also for more general cases, such as when the Lagrangian itself is varied.

If Schwinger's principle is applied specifically to variations which vanish at the initial and final times, the left-hand side (237), and therefore the variation of the action integral, can be equated to zero; since the states $|q_{r1})$ and $|q_{rn})$ are arbitrary, the vanishing of the operator $\int L dt$ is the necessary consequence of the vanishing of its matrix elements. We are thus led to the principle of least action in its correlated operator form. Just as in classical mechanics, this principle in Hamilton's version is tantamount to Lagrange's equations. But, realizing that the commutation relations, which are peculiar to quantum mechanics alone, cannot be derived from the principle of least action as stated here, we have also to examine the variational equation (237) for variations which do not vanish at the termini. The result of varying the action integral in this more general case can be taken over from eq. (53), the derivation of which is valid independent of whether the dynamical variables are c-numbers or operators. Consider, in particular, variations which vanish at the initial time only. Eq. (53) can then be written as

$$\delta(q_{rn}|q_{r1}) = i\left(q_{rn}\left|\frac{\partial L_n}{\partial \dot{q}_{rn}}\delta_n q_r - H_n\delta_n t\right|q_{r1}\right). \quad . \quad (238)$$

The effect of the variation is therefore to apply the unitary operator

$$U = 1 + iG$$

to the conjugate imaginary of the final state, where the infinitesimal operator is given by

$$G = \frac{\partial L}{\partial \dot{q}_r}\delta q_r - H\delta t. \quad . \quad . \quad (239)$$

Alternatively, the final state itself can be regarded as fixed, and we may suppose the operator O acting on it to undergo a corresponding unitary transformation $UOU^{-1} = O + \delta O$, where

$$\delta O = -i[O,G]. \quad . \quad . \quad (240)$$

We are now in the position to determine the commutation relations by simply substituting q_s for O and restricting ourselves to a variation in which only the co-ordinate q_r is altered at time t_n. From (239) and (240) we obtain

$$\delta q_s = -i\left[q_s, \frac{\partial L}{\partial \dot{q}_r}\delta q_r\right]. \qquad . \qquad . \qquad (241)$$

The right-hand side of (241) will reduce to δq_s if and only if

$$\left[q_s, \frac{\partial L}{\partial \dot{q}_r}\right] = i\delta_{rs}. \qquad . \qquad . \qquad (242)$$

The canonical commutation rules are thus a consequence of Schwinger's variational principle.

We can go even further and derive the Heisenberg equations of motion from the same principle, by choosing a variation in which only the time at t_n is changed by the amount δt. Such a variation transforms the final state, which is a simultaneous eigenstate of the operators $q_r(t)$, into a simultaneous eigenstate of the operators $q_r(t + \delta t)$ with the *same*(!) eigenvalues, for we assume the δq_r to be zero. Accordingly, the equivalent change in an operator O is $-(dO/dt)\delta t$. On substituting this value in eq. (240) and using eq. (239) it follows that

$$\frac{dO}{dt} = i[O, H], \qquad . \qquad . \qquad (243)$$

which is Heisenberg's equation for the change in time of an operator. This is not an independent result of Schwinger's theory, since the Heisenberg equations are deducible from Lagrange's equations and the commutation rules, but it demonstrates that consistent answers are reached by applying Schwinger's principle to different types of variation.

From this brief analysis of the alternative Lagrangian formulations of quantum mechanics, it will be appreciated that the work of Feynman and Schwinger represents a notable advance in our knowledge and understanding of the principles underlying the subject. This contention is markedly borne out in quantum field theory, where—in the new approach—the mathematical formalism is manifestly co-variant with respect to Lorentz transformations, provided we adhere to a relativistically invariant Lagrangian. The recent formulations have also been invaluable from a practical point of view; with their aid, Feynman was able to promote considerably our knowledge of quantum electrodynamics as well as to develop approximation methods for solving certain problems, which have proved to be more accurate than any other techniques previously employed.

Attempts have been made to use the Feynman and Schwinger mode of reasoning to attack the most fundamental issue in quantum theory, which is the description of strongly interacting fields. It must be conceded, however, that these endeavours have, as yet, met with little success. The reason may well be—we are wary not to yield to assertive predictions—that the problem will prove to be intrinsically incapable of a solution in the present framework of field theory. Perhaps a revolutionary idea is needed, not so much for the purpose of solving the existing equations, but in order to create a new theory. In the quest for such a conceptual system, the Feynman or Schwinger principle could emerge as a guiding factor of very great significance.

§ 13

Variational Principles in Hydrodynamics

by L. Mittag, M. J. Stephen, and W. Yourgrau

Introduction

THE great success of variational principles in classical mechanics has stimulated many efforts to formulate the laws of hydrodynamics in a similar way. In real fluids exhibiting energy dissipation (on account of thermal conduction or viscosity), as in all systems exhibiting dissipation, the appropriate variational principle takes a form which is dependent upon the particular mechanism of dissipation. Thus variational principles for dissipative systems are in a sense merely reformulations of the equations of motion. These principles do, however, provide methods for handling constraints which one might find difficult to incorporate in the original equations. For dissipative systems it is not always possible to formulate a variational principle. These remarks do not apply to ideal (i.e., non-dissipative) fluids, where one can write a variational principle without *a priori* knowledge of the equations of motion. It is surprising that a correct variational principle for hydrodynamics has only recently been obtained by Herivel.[43] This principle, in the case of isentropic flow, leads to irrotational motion of the fluid. The necessary extension, to include the case of rotational motion, has been given by Lin.[44] The variational principles discussed below are closely related to those of Herivel and Lin.

In hydrodynamics we are concerned with the motion of liquids and gases. The phenomena considered are macroscopic and the atomic or molecular nature of the fluid is neglected. Thus the fluid is regarded as a continuous medium. This implies that any small volume element is always supposed to be so large that it still contains a large number of molecules. Accordingly, when we speak of infinitesimal volume elements we mean that they are small compared with the volume of the system but still sufficiently large to contain very many molecules.

Of the two representations normally employed in the study of hydrodynamics, the Lagrangian form is more closely related to classical mechanics, but the Eulerian form is better known. In the Eulerian description, a one-component moving fluid is described by five quantities: the components of the fluid velocity $\mathbf{v}(\mathbf{x}, t)$ and two thermodynamic quantities. These two are most commonly taken to be the pressure $P(\mathbf{x}, t)$ and the density $\rho(\mathbf{x}, t)$. From P and ρ all other thermodynamic quantities can be determined, provided the equation of state is known. This is a typical field theory, in which the observables are given as a function of the position vector $\mathbf{x}$ and the time t.

The function $\mathbf{v}(\mathbf{x}, t)$, furnishing the velocity of the fluid at a given point $\mathbf{x}$ at a time t, refers to fixed points in space and not to definite particles of the fluid. The acceleration of a particular fluid particle is then given by the total derivative

$$\frac{d\mathbf{v}}{dt} = \frac{\partial \mathbf{v}}{\partial t} + \mathbf{v} \cdot \mathrm{grad}\ \mathbf{v}.$$

In the Lagrangian description, a particular fluid particle is selected and its motion followed in time. Let its position initially (at $t = 0$) be $\mathbf{a}$ and its position at a later time t be $\mathbf{x}$. Then we consider $\mathbf{x}(\mathbf{a}, t)$ as a function of the initial position $\mathbf{a}$ and the time t. Clearly, $\mathbf{x}(\mathbf{a}, 0) = \mathbf{a}$. In the case of periodic motion, all that is needed is the position of the fluid particle at some time during the cycle, since all particles can be traced back in time. In the continuum approximation, valid for non-turbulent hydrodynamics, $\mathbf{x}(\mathbf{a}, t)$ is a differentiable vector field in all its arguments. The velocity of the fluid particle is then given by $\mathbf{v} = (\partial \mathbf{x}/\partial t)_{\mathbf{a}}$ and the acceleration by $(\partial^2 \mathbf{x}/\partial t^2)_{\mathbf{a}}$.

Before we can apply a variational principle to determine the equations of motion of a fluid, it is necessary to prescribe the constraints; for an ideal fluid we must include the following two:

(a) *Conservation of Mass.* Consider the mass contained in a small volume $d^3\mathbf{x}$ of the fluid surrounding a fluid particle at $\mathbf{x}(\mathbf{a}, t)$. This defines the density $\rho(\mathbf{x}, t)$. Since this mass was contained initially in a similar volume about the point $\mathbf{a}$, we must have

$$\rho(\mathbf{x}, t)d^3\mathbf{x} = \rho(\mathbf{a}, 0)d^3\mathbf{a}.$$

If we let J be the Jacobian $J = \dfrac{\partial(x_1, x_2, x_3)}{\partial(a_1, a_2, a_3)}$, then the conservation of mass is given by

$$\rho(\mathbf{x}, t)J = \rho(\mathbf{a}, 0) = \rho_0. \qquad . \qquad . \qquad (244)$$

Specially for an incompressible fluid, $J = 1$.

The constraint (244) may be compared with the equation of continuity, which is its counterpart in the Eulerian formalism. Here we focus our attention on a small volume ΔV at fixed $\mathbf{x}$ at time t. The rate of increase of the mass ΔM contained in this volume element must be equal to the mass of fluid flowing inward across the boundary σ of the volume element. Thus,

$$\frac{d(\Delta M)}{dt} \equiv \int_{\Delta V} \frac{\partial \rho}{\partial t}\, d^3\mathbf{x}$$

$$= -\int_\sigma \rho \mathbf{v} \cdot d\boldsymbol{\sigma} = -\int_{\Delta V} \operatorname{div}(\rho\mathbf{v}) d^3\mathbf{x}.$$

As the volume element ΔV is arbitrary, the conservation of mass is given by

$$\frac{\partial \rho}{\partial t} + \operatorname{div}(\rho\mathbf{v}) = 0. \qquad . \qquad . \qquad (245)$$

The condition for an incompressible fluid is now div $\mathbf{v} = 0$.

(b) *Conservation of Entropy.* In real fluids, energy dissipation occurs as a consequence of viscosity, thermal conduction, etc. But since we are studying an ideal fluid, such irreversible processes are absent, and the fluid motion will take place without any change in entropy. Let $s(\mathbf{x}, t)$ be the entropy per unit mass of the fluid at the point $\mathbf{x}$ at time t. Then, by considering the entropy contained in a small volume element at $\mathbf{x}$, one can derive one form of the conservation of entropy:

$$\rho(\mathbf{x}, t)s(\mathbf{x}, t)J = \rho(\mathbf{a}, 0)s(\mathbf{a}, 0),$$

or, because $\rho(\mathbf{x}, t)J = \rho(\mathbf{a}, 0)$,

$$s(\mathbf{x}, t) = s(\mathbf{a}, 0) = s_0. \qquad . \qquad . \qquad (246)$$

In the Eulerian form, the conservation of entropy is expressed by

$$\frac{\partial}{\partial t}(\rho s) + \operatorname{div}(\rho s\mathbf{v}) = 0, . \qquad . \qquad . \qquad (247)$$

which, on using (245), can also be written as

$$\frac{\partial s}{\partial t} + \mathbf{v} \cdot \operatorname{grad} s = 0. \qquad (248)$$

When the motion of the fluid obeys (246), (247), or (248), it is said to be isentropic.

Lagrangian Variational Principle

With reference to a discrete dynamical system, in which we may ignore thermodynamics, Hamilton's principle for a conservative system is

$$\delta I = \delta \int_{t_0}^{t_1} dt(K - V) = 0, \qquad (249)$$

where K is the kinetic energy† and V the potential energy.

A generalization of this equation (see Herivel[43]) to a continuous system, in which we cannot ignore thermodynamics, is to redefine V as the sum of the internal energy and the external potential energy. Thus the kinetic and potential energies of the fluid are, if we adopt the convention that repeated indices are summed over,

$$K = \int \tfrac{1}{2}\rho \left(\frac{\partial x_i}{\partial t}\right)^2 d^3\mathbf{x} = \int \tfrac{1}{2}\rho_0 \left(\frac{\partial x_i}{\partial t}\right)^2 d^3\mathbf{a} \qquad (250)$$

and

$$V = \int \rho(e + U)d^3\mathbf{x} = \int \rho_0(e + U)d^3\mathbf{a}. \qquad (251)$$

We assume here that the external forces are derivable from a potential U (although this is not strictly necessary) and that the internal energy e is a function of the coordinates only indirectly through the density and entropy: $e = e(\rho, s)$. Both U and e are measured per unit mass, as is customary in hydrodynamics. The thermodynamic derivatives are then

$$P = \rho^2 \left(\partial e/\partial \rho\right)_s \quad \text{and} \quad T = (\partial e/\partial s)_\rho, \qquad (252)$$

where P denotes the pressure and T the temperature.

We are now able to formulate a variational principle for hydrodynamics: δI, given by equation (249), should be minimized, subject to the constraints (244) and (246). It is simplest to introduce the constraints by way of Lagrange multipliers. We consider

$$\delta \int_{t_0}^{t_1} dt \int d^3\mathbf{a} \left[\frac{1}{2}\rho_0 \left(\frac{\partial x_i}{\partial t}\right)^2 - \rho_0(e + U) \right.$$
$$\left. + \alpha(\rho J - \rho_0) + \beta(s - s_0)\right] = 0, \qquad (253)$$

†We use the symbol K for kinetic energy, instead of T, as designated in §5. The reason is that in this chapter T denotes temperature.

where K and V in (249) have been replaced by the expression (250) and (251), and $\alpha(\mathbf{x}, t)$ and $\beta(\mathbf{x}, t)$ are Lagrange multipliers. The use of these multipliers enables us to take variations with respect to the variables ρ, s and $\mathbf{x}$ independently in (253). Thus, variations with respect to ρ and s yield, on using (252), respectively

$$\rho\alpha = P \quad \text{and} \quad \beta = \rho_0 T. \qquad . \qquad . \qquad (254)$$

Hence the density constraint is related to the pressure, and the entropy constraint to the temperature.

To obtain the variation in $\mathbf{x}$, we note that both spatial and time derivatives of $\mathbf{x}$, i.e., J and all $\partial x_i/\partial t$, appear in the integrand of (253). The Euler–Lagrange equation is therefore

$$\frac{\partial}{\partial t} \frac{\partial l}{\partial \left(\dfrac{\partial x_i}{\partial t}\right)} + \frac{\partial}{\partial a_j} \frac{\partial l}{\partial \left(\dfrac{\partial x_i}{\partial a_j}\right)} - \frac{\partial l}{\partial x_i} = 0,$$

l signifying the integrand. When the operations indicated in this equation are carried out, one gets

$$\rho_0 \frac{\partial^2 x_i}{\partial t^2} + \frac{\partial}{\partial a_j} (\rho\alpha J_{ij}) + \rho_0 \frac{\partial U}{\partial x_i} = 0, . \qquad . \qquad (255)$$

where J_{ij} is the (i, j) minor of the determinant J. Since $\dfrac{\partial}{\partial a_j} J_{ij} = 0$ and $\rho\alpha = P$, the second term in (255) becomes

$$\frac{\partial}{\partial a_j} (\rho\alpha J_{ij}) = \frac{\partial P}{\partial a_j} J_{ij} = \frac{\partial P}{\partial x_k} \frac{\partial x_k}{\partial a_j} J_{ij}. \qquad . \qquad (256)$$

But the expansion of a determinant in terms of its minors satisfies $\dfrac{\partial x_k}{\partial a_j} J_{ij} = J\delta_{ik}$; on substituting this result in (256) together with $J = \rho_0/\rho$, one obtains from (255) the Lagrangian equation of motion

$$\left(\frac{\partial^2 x_i}{\partial t^2}\right)_{\mathbf{a}} + \frac{1}{\rho} \frac{\partial P}{\partial x_i} = -\frac{\partial U}{\partial x_i} . \qquad . \qquad . \qquad (257)$$

We note that the entropy constraint had little effect on the equation of motion (257). One could have ignored it and considered the internal energy as a function of the density only. If the external forces are not derivable from a potential, we should consider, in place of (249),

$$\int_{t_0}^{t_1} dt \left[\delta K - \int d^3\mathbf{a}(\delta(\rho_0 e) - \rho_0 F_i \delta x_i) \right] = 0,$$

F denoting the vector sum of the forces acting on the fluid particle at **x**.

The equation of motion (257) may be put in a more familiar form. Since $v_i = (\partial x_i / \partial t)_a$, it follows that

$$\left(\frac{\partial v_i}{\partial t}\right)_a = \left(\frac{\partial v_i}{\partial t}\right)_a + \frac{\partial v_i}{\partial x_k}\left(\frac{\partial x_k}{\partial t}\right)_a = \left(\frac{\partial v_i}{\partial t}\right)_x + v_k\left(\frac{\partial v_i}{\partial x_k}\right).$$

Substitution of this result in (257) yields Euler's equation

$$\frac{\partial \mathbf{v}}{\partial t} + \mathbf{v}\,.\,\text{grad }\mathbf{v} + \frac{1}{\rho}\,\text{grad }P = \mathbf{F}. \qquad . \qquad (258)$$

Eulerian Variational Principle

Attempts to derive Euler's equation from a variational principle have suffered from the defect (see Eckart[45] and Herivel[43]) that in the situation where the entropy is constant the flow is restricted to be irrotational. Lin[44] has pointed out that the introduction of Lagrangian coordinates as constraints resolves this difficulty. Thus, given $\mathbf{x}(\mathbf{a}, t)$, one may solve for $\mathbf{a}(\mathbf{x}, t)$. Since the a_i are independent of time, they satisfy

$$\frac{da_i}{dt} = \left(\frac{\partial a_i}{\partial t}\right)_x + \frac{\partial a_i}{\partial x_j}\frac{\partial x_j}{\partial t} = \left(\frac{\partial a_i}{\partial t}\right)_x + v_j\frac{\partial a_i}{\partial x_j} = 0. \quad . \quad (259)$$

This is a further constraint that must be imposed on the Eulerian form of the variational principle. It is a consequence of the fact that fluid particles are distinguishable owing to their different initial positions.

When we introduce Lagrange multipliers $-\phi$ and $\rho\gamma_i$ for the continuity equation (245) and the Lagrange coordinate equation (259), respectively, the variational principle takes the form

$$\delta \int_{t_0}^{t_1} dt \int d^3\mathbf{x} \left[\tfrac{1}{2}\rho\mathbf{v}^2 - \rho(e + U) - \phi\left(\frac{\partial \rho}{\partial t} + \text{div }\rho\mathbf{v}\right) \right.$$
$$\left. + \rho\gamma_i\left(\frac{\partial a_i}{\partial t} + v_j\frac{\partial a_i}{\partial x_j}\right) \right] = 0. \qquad . \qquad (260)$$

The entropy constraint has been omitted as it has no effect on the result. The internal energy e will be regarded as a function of ρ alone. Provided we restrict ourselves in (260) to variations that vanish at the boundaries of the region of integration, the variations with respect to a_i, v_i, and ρ are seen to produce the following equations:

$$\frac{\partial}{\partial t}(\rho \gamma_i) + \text{div}(\rho \mathbf{v} \gamma_i) = 0,$$

or
$$\frac{\partial \gamma_i}{\partial t} + \mathbf{v} \cdot \text{grad } \gamma_i = 0, \quad . \quad (261)$$

$$v_i + \frac{\partial \phi}{\partial x_i} + \gamma_j \frac{\partial a_j}{\partial x_i} = 0, \quad . \quad (262)$$

$$\tfrac{1}{2}v^2 - \frac{\partial}{\partial \rho}(\rho e) - U + \frac{\partial \phi}{\partial t} + \mathbf{v} \cdot \text{grad } \phi = 0. \quad . \quad (263)$$

If we subtract the gradient of (263) from the time derivative of (262) and again use (262) to eliminate ϕ, we find

$$\frac{\partial v_i}{\partial t} + \frac{\partial}{\partial x_i}\left[\frac{\partial}{\partial \rho}(\rho e) + \frac{1}{2}v^2\right] + \frac{\partial}{\partial t}\left(\gamma_j \frac{\partial a_j}{\partial x_i}\right)$$
$$+ \frac{\partial}{\partial x_i}(\gamma_j \mathbf{v} \cdot \text{grad } a_j) = -\frac{\partial U}{\partial x_i}. \quad . \quad . \quad (264)$$

By virtue of (252) and since $e = e(\rho)$, the second term on the left-hand side can be written as

$$\frac{\partial}{\partial x_i}\frac{\partial}{\partial \rho}(\rho e) = \frac{1}{\rho}\frac{\partial P}{\partial x_i}. \quad . \quad . \quad . \quad (265)$$

In the last two terms on the left-hand side of (264), eqs. (259) and (261) can be invoked to eliminate the time derivatives; with the aid of (262), one gets

$$\frac{\partial}{\partial t}\left(\gamma_j \frac{\partial a_j}{\partial x_i}\right) + \frac{\partial}{\partial x_i}(\gamma_j \mathbf{v} \cdot \text{grad } a_j) = v_k\left(\frac{\partial \gamma_j}{\partial x_i}\frac{\partial a_j}{\partial x_k} - \frac{\partial \gamma_j}{\partial x_k}\frac{\partial a_j}{\partial x_i}\right)$$
$$= -(\mathbf{v} \times \text{curl } \mathbf{v})_i. \quad . \quad (266)$$

Finally, substituting these results in (264), we obtain Euler's equation:

$$\frac{\partial \mathbf{v}}{\partial t} + \text{grad}(\tfrac{1}{2}v^2) - \mathbf{v} \times \text{curl } \mathbf{v} + \frac{1}{\rho}\text{grad } P = \mathbf{F}. \quad (267)$$

Application of the vector identity

$$\text{grad } \tfrac{1}{2}v^2 - \mathbf{v} \times \text{curl } \mathbf{v} = (\mathbf{v} \cdot \text{grad})\mathbf{v}$$

allows this equation to be cast in the usual form

$$\frac{\partial \mathbf{v}}{\partial t} + (\mathbf{v} \cdot \mathrm{grad})\mathbf{v} + \frac{1}{\rho}\,\mathrm{grad}\,P = \mathbf{F}. \qquad (268)$$

The variational principle (260) may be simplified. Thus eq. (261) for the Lagrange multiplier γ_i is in the form of a conservation law and is related to the vorticity. If we define $\omega = \mathrm{curl}\,\mathbf{v}$ and find solutions of the differential equations

$$\frac{dx_1}{\omega_1} = \frac{dx_2}{\omega_2} = \frac{dx_3}{\omega_3}$$

in the form $\alpha(\mathbf{x}, t) = \mathrm{const.}$ and $\beta(\mathbf{x}, t) = \mathrm{const.}$, it can be shown that there exist two functions $\lambda(\alpha, \beta)$, $\mu(\alpha, \beta)$ such that the velocity is furnished by (see Lamb[46])

$$\mathbf{v} + \mathrm{grad}\,\phi + \lambda\,\mathrm{grad}\,\mu = 0$$

instead of (262). Both λ and μ satisfy the same equation as a_i and γ_i (see (259) and (261)), that is,

$$\frac{\partial \lambda}{\partial t} + \mathbf{v} \cdot \mathrm{grad}\,\lambda = 0 \quad \text{and} \quad \frac{\partial \mu}{\partial t} + \mathbf{v} \cdot \mathrm{grad}\,\mu = 0. \qquad (269)$$

This transformation, due to Clebsch, supplies the vortex lines as the intersections of the surfaces $\lambda = \mathrm{constant}$ and $\mu = \mathrm{constant}$. Thus it is possible to repeat the variational calculation employing only one of the vortex conservation laws. One need merely examine

$$\delta \int_{t_0}^{t_1} dt \int d^3\mathbf{x} \left[\tfrac{1}{2}\rho\mathbf{v}^2 - \rho(e + U) - \phi\left(\frac{\partial \rho}{\partial t} + \mathrm{div}\,(\rho\mathbf{v})\right) + \right.$$

$$\left. \rho\lambda\left(\frac{\partial \mu}{\partial t} + \mathbf{v} \cdot \mathrm{grad}\,\mu\right)\right] = 0$$

to obtain Euler's equation including vorticity. Indeed in this form it is easier to set up a canonical transformation for a quantum field theory.

Conservation of Momentum and Energy; Dissipative Processes

We have not yet exhausted our variational principle. Let us consider systems with no external forces. It is well known that invariance under space and time translations imply conservation of momentum and energy, respectively. By restricting the variations in (260) to vanish at the boundaries, we have been able to derive the equations of motion. If now, adopting these equations, one allows finite variations at the boundaries, the energy momentum tensor can be derived. Omitting details we obtain

$$\frac{\partial}{\partial t}(\rho v_i) + \frac{\partial}{\partial x_j}(\rho v_i v_j + P\delta_{ij}) = 0 \quad . \quad (270)$$

and

$$\frac{\partial}{\partial t}(\rho e + \tfrac{1}{2}\rho v^2) + \text{div}\,((\rho e + P + \tfrac{1}{2}\rho v^2)\mathbf{v}) = 0, \quad . \quad (271)$$

with $\rho v_i v_j + P\delta_{ij}$ as the pressure–momentum or stress tensor

and $(\rho e + P + \tfrac{1}{2}\rho v^2)\mathbf{v}$ as the energy flux density.

The interpretation of the first terms in these equations as momentum and energy density is clear. Alternatively, the conservation laws (270) and (271) may be derived in a straightforward manner from Euler's equation (268), the equations of continuity (245) and (247), and thermodynamic relations.

In the presence of dissipative processes, which have until now been ignored, momentum and energy are still conserved and, therefore, equations (270) and (271) must essentially be preserved. We first examine the conservation of momentum in the presence of dissipative processes.

Consider the rate of change of the momentum $\Delta\mathbf{p}$ contained in a small volume Δv,

$$\frac{d}{dt}(\Delta p) = \int_{\Delta V} \frac{\partial(\rho\mathbf{v})}{\partial t}\,d^3\mathbf{x}.$$

This must be related to the flux of momentum across the boundary $\boldsymbol{\sigma}$ of Δv,

$$\frac{d}{dt}(\Delta p_i) = -\int_\sigma \Pi'_{ij}d\sigma_j,$$

where Π'_{ij} is of necessity a second-rank tensor. If we define Π_{ij} as the pressure–momentum tensor in the absence of dissipative processes less Π'_{ij}, so that

$$\rho v_i v_j + P\delta_{ij} - \Pi'_{ji} = \Pi_{ij},$$

the foregoing equation gives rise to the momentum conservation relation

$$\frac{\partial}{\partial t}(\rho v_i) + \frac{\partial}{\partial x}(\rho v_i v_j + P\delta_{ij} - \Pi_{ij}) = 0. \quad . \quad (272)$$

When combined with the continuity equation (245), this leads to the Navier–Stokes equation in the form

$$\frac{\partial v_i}{\partial t} + \mathbf{v} \cdot \operatorname{grad} v_i + \frac{1}{\rho}\frac{\partial P}{\partial x_i} = \frac{1}{\rho}\frac{\partial \Pi_{ij}}{\partial x_j}, \qquad \qquad (273)$$

where Π_{ij} has not yet been specified.

We turn now to energy conservation. Let us multiply (273) by ρv_i and then add $\tfrac{1}{2}v^2\left(\dfrac{\partial \rho}{\partial t} + \operatorname{div}(\rho\mathbf{v})\right) = 0$ to get

$$\frac{\partial}{\partial t}\left(\tfrac{1}{2}\rho v^2\right) + \operatorname{div}\left(\tfrac{1}{2}\rho v^2\mathbf{v}\right) + \mathbf{v} \cdot \operatorname{grad} P = v_i\frac{\partial \Pi_{ij}}{\partial x_j}. \qquad (274)$$

Add to this the equation

$$\frac{\partial}{\partial t}(\rho e) + \operatorname{div}((\rho e + P)\mathbf{v}) - \mathbf{v} \cdot \operatorname{grad} P$$
$$= \rho T\left(\frac{\partial s}{\partial t} + \mathbf{v} \cdot \operatorname{grad} s\right), \qquad (275)$$

which is easily derived from the continuity equation (245) and the thermodynamic relation $de = (T ds + P/\rho^2)d\rho$. The result is

$$\frac{\partial}{\partial t}(\rho e + \tfrac{1}{2}\rho v^2) + \operatorname{div}\left[(\rho e + P + \tfrac{1}{2}\rho v^2)\mathbf{v} - \mathbf{v} \cdot \mathbf{\Pi}\right]$$
$$= \rho T\left(\frac{\partial s}{\partial t} + \mathbf{v} \cdot \operatorname{grad} s\right) - \Pi_{ij}\frac{\partial v_i}{\partial x_j}. \qquad \qquad (276)$$

If we examine the energy ΔE contained in a small volume ΔV, we find, with $\mathbf{Q}$ denoting the energy flux density,

$$\frac{d}{dt}(\Delta E) = \int_{\Delta V}\frac{\partial}{\partial t}(\rho e + \tfrac{1}{2}\rho v^2)d^3\mathbf{x} = -\int_{\sigma}\mathbf{Q} \cdot d\boldsymbol{\sigma},$$

since energy is conserved. Further, let $\mathbf{q}$ be the difference between $\mathbf{Q}$ and the energy flux excluding thermal conduction, i.e.

$$\mathbf{q} = \mathbf{Q} - (\tfrac{1}{2}\rho v^2 + \rho e + P)\mathbf{v} + \mathbf{v} \cdot \mathbf{\Pi}, \qquad (277)$$

then the energy conservation relation will read

$$\frac{\partial}{\partial t}(\rho e + \tfrac{1}{2}\rho v^2) + \operatorname{div}\left[(\rho e + P + \tfrac{1}{2}\rho v^2)\mathbf{v} - \mathbf{v} \cdot \mathbf{\Pi} + \mathbf{q}\right] = 0.$$
$$\qquad \qquad \qquad (278)$$

According to (276), this implies that

$$T\left[\frac{\partial(\rho s)}{\partial t} + \text{grad}\,(\rho s\mathbf{v})\right] + \text{div}\,\mathbf{q} - \Pi_{ij}\frac{\partial v_i}{\partial x_j} = 0. \qquad (279)$$

Now, as $1/T\,\text{div}\,\mathbf{q} = \text{div}\,(\mathbf{q}/T) - \mathbf{q}\,.\,\text{grad}\,(1/T)$, eq. (279) can be put in the form

$$\frac{\partial(\rho s)}{\partial t} + \text{div}\left(\rho s\mathbf{v} + \frac{\mathbf{q}}{T}\right) = \mathbf{q}\,.\,\text{grad}\,\frac{1}{T} + \frac{1}{T}\,\Pi_{ij}\frac{\partial v_i}{\partial x_j}. \qquad (280)$$

Since $\mathbf{q}/T$ is the entropy flux due to thermal conduction, $\rho s\mathbf{v} + \mathbf{q}/T$, appearing on the left, is entropy flux. The right-hand side of (280) is the entropy production per unit volume due to irreversible processes. The entropy production for the total fluid is

$$\dot{S} = \int \frac{\partial(\partial s)}{\partial t}\,d^3\mathbf{x} = \int d^3\mathbf{x}\left(\mathbf{q}\,.\,\text{grad}\,\frac{1}{T} + \frac{1}{T}\,\Pi_{ij}\frac{\partial v_i}{\partial x_j}\right) \qquad (281)$$

in terms of the fluxes and gradients.

We have still to determine the fluxes $\mathbf{q}$ and Π_{ij} of heat and momentum. These fluxes must vanish if the gradients of the corresponding thermodynamic variables T and $\mathbf{v}$ do, so that if one expands a flux in terms of these gradients, there can be no term independent of the gradient. Moreover, cross-terms must be absent, since viscosity and thermal conduction are essentially independent processes—although both are related to atomic or molecular diffusion. Finally, according to the second law of thermodynamics, $\dot{S}$ must be positive. All these requirements can be met by the posits

$$\mathbf{q} = k\,\text{grad}\,\frac{1}{T}$$

and
$$\Pi_{ij} = \eta\left(\frac{\partial v_i}{\partial x_j} + \frac{\partial v_j}{\partial x_i}\right) + \zeta\,(\text{div}\,\mathbf{v})\delta_{ij}, . \qquad . \qquad (282)$$

where k, η, and ζ are the coefficients of thermal conductivity, viscosity, and bulk viscosity, all of which are necessarily positive.

Hydrodynamics of Superfluids

At very low temperatures quantum effects become important in the properties of fluids. There is only one substance which can remain fluid down to absolute zero, namely helium; all other substances solidify before quantum effects become noticeable. At a temperature of about $2\cdot18^\circ$ K and pressures below 25 atmospheres, the isotope He⁴ makes a phase transition. The new phase of liquid helium is generally referred to as helium II; it has some remarkable properties, the most interesting being superfluidity. This phenomenon, discovered by P. L. Kapitza in 1938, is the capacity of He II to flow without viscosity through capillaries and gaps.

The hydrodynamics of He II is based on the observation that it behaves as if it were a mixture of two interpenetrating fluids. One fluid, called the superfluid, is believed to be a manifestation of a macroscopic quantum state resulting from a Bose–Einstein condensation. The superfluid carries no entropy, has an irrotational velocity field, and flows without viscosity. The reason for this inviscid behaviour is presumably the absence of other low-lying flow states into which the fluid can scatter. The other fluid, known as the normal fluid, is the aggregate of all excitations or quasi-particles that, at finite temperatures, have been excited out of the Bose condensate. The velocity field of the normal fluid is then the average velocity of the excitations computed with the aid of the appropriate temperature-dependent distribution function. Irreversible processes similar to those occurring in ordinary fluids can take place in the normal fluid. We owe this picture of He II mainly to the work of Landau,[47] London,[48] and Tisza.[49]

The equations of motion of the two-fluid system were first derived by Landau by an intuitive method. It then became of interest to rederive Landau's results by different methods, in particular by variational principles. The first variational principle, due to Zilsel,[50] suffered from the defect that, when applied to an ordinary fluid, it led to potential flow only. One method for obtaining the two-fluid equations that is not based on variational principles is of considerable interest. It is due to Khalatnikov who showed that the flow equations can be uniquely determined simply from the requirements imposed by Galilean invariance and by the necessary conservation laws together with the properties of the superfluid discussed above. Finally, Lin,[44] who assumed only that the velocities of mass and entropy flows differed, was able to derive a set of two-fluid equations from a variational principle without the need for an irrotational velocity field. However, the existence of a flow that carries mass but not entropy would seem to imply that the velocity field is irrotational. Thus, if the flow carries no entropy, a single quantum state is a possible description of the flow state, and a single quantum state is of necessity always associated with an irrotational velocity field. The variational principle discussed here is most closely related to that of Lin.

We begin by discussing the form of the conservation relations appropriate to the two-fluid model of He II. The total mass flux $\mathbf{j}$ is the sum of the fluxes of the superfluid and normal flows:

$$\mathbf{j} = \rho_s \mathbf{v}_s + \rho_n \mathbf{v}_n, \qquad \cdot \qquad \cdot \qquad \cdot \qquad (283)$$

where $\mathbf{v}_s$ and $\mathbf{v}_n$ are the velocity fields of the superfluid and normal flows. The coefficients ρ_s and ρ_n are called the superfluid and normal densities, respectively, and are functions of the temperature; ρ_n vanishes at $0°$ K, where the fluid becomes entirely superfluid, while ρ_s vanishes at the phase transition, when the fluid becomes wholly normal. The total mass density is given by

$$\rho = \rho_s + \rho_n. \qquad \cdot \qquad \cdot \qquad \cdot \qquad (284)$$

Conservation of mass is symbolized by

$$\frac{\partial \rho}{\partial t} + \operatorname{div} \mathbf{j} = 0. \qquad \cdot \qquad \cdot \qquad \cdot \qquad (285)$$

Because entropy is associated only with the normal fluid, its conservation is expressed by

$$\frac{\partial}{\partial t}(\rho s) + \operatorname{div}(\rho s \mathbf{v}_n) = 0. \qquad \cdot \qquad \cdot \qquad (286)$$

In the absence of dissipative processes, the two-fluid equations of Landau can now be derived from the following variational principle:

$$\delta \int_{t_0}^{t_1} dt \int d^3\mathbf{x} \left[\left(\tfrac{1}{2}\rho_s v_s{}^2 + \tfrac{1}{2}\rho_n v_n{}^2\right) - \rho e - \phi \left(\frac{\partial \rho}{\partial t} + \operatorname{div} \mathbf{j}\right) \right.$$
$$\left. - \beta \left(\frac{\partial(\rho s)}{\partial t} + \operatorname{div}(\rho s \mathbf{v}_n)\right) - \lambda \left(\frac{\partial(\rho\mu)}{\partial t} + \operatorname{div}(\rho\mu\mathbf{v}_n)\right) \right] = 0.$$
$$\cdot \qquad \cdot \qquad \cdot \qquad (287)$$

The first two terms represent the kinetic energy of the fluid. The potential energy due to external forces has been neglected for simplicity, and thus the potential energy per unit mass is given by the internal energy $e(\rho, s, \alpha)$, where $\alpha = \rho_n/\rho$ is the normal-fluid concentration. The conservation of vorticity is given by the last term in (287), and the vortex density must move with the normal fluid owing to the irrotational nature of the superfluid. We obtain the following equations by equating to zero the variations in (287) with respect to the variables s, μ, $\mathbf{v}_s$, $\mathbf{v}_n$, ρ_s, and ρ_n:

$$\frac{\partial \beta}{\partial t} + \mathbf{v}_n \cdot \operatorname{grad} \beta = T, \qquad (288)$$

$$\frac{\partial \lambda}{\partial t} + \mathbf{v}_n \cdot \operatorname{grad} \lambda = 0, \qquad (289)$$

$$\mathbf{v}_s + \operatorname{grad} \phi = 0, \qquad (290)$$

$$\rho_n(\mathbf{v}_n + \operatorname{grad} \phi) + \rho(s \operatorname{grad} \beta + \mu \operatorname{grad} \lambda) = 0, \qquad (291)$$

$$\tfrac{1}{2}v_s{}^2 - e - \frac{P}{\rho} + \frac{\partial e}{\partial \alpha}\frac{\rho_n}{\rho} + \frac{\partial \phi}{\partial t} + \mathbf{v}_s \cdot \operatorname{grad} \phi + sT = 0, \qquad (292)$$

$$\tfrac{1}{2}v_n{}^2 - e - \frac{P}{\rho} - \frac{\partial e}{\partial \alpha}\frac{\rho s}{\rho} + \frac{\partial \phi}{\partial t} + \mathbf{v}_n \cdot \operatorname{grad} \phi + sT = 0. \qquad (293)$$

Let it first be noted that $\mathbf{v}_s$ is irrotational while $\mathbf{v}_n$ is not. Eliminating $\partial \phi / \partial t$ from (292) and (293) and substituting grad $\phi = -\mathbf{v}_s$, one finds

$$\left(\frac{\partial e}{\partial \alpha}\right)_{\rho,s} = \tfrac{1}{2}(\mathbf{v}_n - \mathbf{v}_s)^2. \qquad (294)$$

We shall discuss this result further below, but here we observe that in the absence of any internal convection, i.e., when $\mathbf{v}_n = \mathbf{v}_s$, this equation becomes

$$\left(\frac{\partial e}{\partial \alpha}\right)_{\rho,s} = 0.$$

This is the usual thermodynamic condition for equilibrium of a two-phase system where the particles making up the two phases are interchangeable. Equation (294) then represents a shift in the equilibrium concentration when internal convection is present.

The chemical potential per unit mass, g, is equal to

$$e + \frac{P}{\rho} - Ts,$$

so that, with the help of (290) and (294), eq. (292) becomes

$$\tfrac{1}{2}v_s{}^2 + g - \tfrac{1}{2}\alpha(\mathbf{v}_n - \mathbf{v}_s)^2 - \frac{\partial \phi}{\partial t} = 0. \qquad (295)$$

From (290) the equation of motion of the superfluid is

$$\frac{\partial \mathbf{v}_s}{\partial t} + \operatorname{grad} \frac{\partial \phi}{\partial t} = 0;$$

and elimination of $\partial\phi/\partial t$ by means of (295) gives finally

$$\frac{\partial \mathbf{v}_s}{\partial t} + \text{grad} \left(g + \tfrac{1}{2}v_s{}^2 - \tfrac{1}{2}\alpha(\mathbf{v}_n - \mathbf{v}_s)^2 \right) = 0. \quad . \quad (296)$$

An alternative form of this equation is obtained on using the thermodynamic relation

$$dg = \frac{1}{\rho}\, dP - s\, dT + \left(\frac{\partial e}{\partial\alpha}\right)_{\rho,s} d\alpha, \quad . \quad . \quad (297)$$

so that

$$\left(\frac{\partial g}{\partial\alpha}\right)_{P,T} = \left(\frac{\partial e}{\partial\alpha}\right)_{\rho,s} = \tfrac{1}{2}(\mathbf{v}_n - \mathbf{v}_s)^2.$$

By the last result, (296) can be put in the form

$$\frac{\partial \mathbf{v}_s}{\partial t} + \frac{1}{\rho}\,\text{grad}\,P - s\,\text{grad}\,T + \text{grad}\,(\tfrac{1}{2}v_s{}^2) - \tfrac{1}{2}\alpha\,\text{grad}\,(\mathbf{v}_n - \mathbf{v}_s)^2$$
$$= 0. \quad . \quad (298)$$

The stationary solution of this equation in the absence of flow, $\text{grad}\,P = \rho s\,\text{grad}\,T$, is the well-known thermo-mechanical effect; thus, if two vessels containing He II at different temperatures are connected by a narrow capillary, superfluid flow will take place until the stated pressure difference is established.

The corresponding equation for $\mathbf{v}_n$ can be derived in a similar manner to (298). Omitting details, the result is

$$\frac{\partial \mathbf{v}_n}{\partial t} + \mathbf{v}_n \cdot \text{grad}\,\mathbf{v}_n + \frac{1}{\rho}\,\text{grad}\,P + \frac{\rho_s}{\rho_n}\,s\,\text{grad}\,T$$
$$+ \tfrac{1}{2}(1 - \alpha)\,\text{grad}\,(\mathbf{v}_n - \mathbf{v}_s)^2 + \frac{1}{\rho_n}\,(\mathbf{v}_n - \mathbf{v}_s)f = 0, \quad (299)$$

in which f is the rate of conversion of superfluid to normal fluid defined by

$$\frac{\partial\rho_n}{\partial t} + \text{div}\,(\rho_n\mathbf{v}_n) = f$$

and

$$\frac{\partial\rho_s}{\partial t} + \text{div}\,(\rho_s\mathbf{v}_s) = -f. \quad . \quad . \quad (300)$$

It is interesting to note that the source function f only appears in the equation for the normal fluid velocity (299). The physical reason for this is that the superfluid velocity is the velocity of the background Bose condensate, whereas the normal fluid velocity is a weighted average over the velocities of the thermal excitations present.

The momentum conservation equation can be deduced from the above equations and takes the particularly simple form

$$\frac{\partial \mathbf{j}}{\partial t} + \text{div } (\rho_n \mathbf{v}_n \mathbf{v}_n + \rho_s \mathbf{v}_s \mathbf{v}_s + P\mathbf{I}) = 0, \qquad . \qquad (301)$$

where $\mathbf{I}$ is the unit dyadic.

We may now also consider the changes due to irreversible processes. A detailed analysis has been carried out by Khalatnikov[51] who has demonstrated that there are five independent dissipation coefficients (instead of the three coefficients for an ordinary fluid). Of these coefficients, one is the thermal conductivity k and one is the first viscosity η. They are entirely analogous to the corresponding dissipative coefficients of an ordinary fluid. It is found, moreover, that the pressure momentum tensor in (301) and the quantity whose gradient appears in (296) involve further viscous terms proportional to div $\mathbf{v}_n$ and div $\rho_s(\mathbf{v}_s - \mathbf{v}_n)$. The irrotational nature of the superfluid flow is preserved in the case of dissipation.

The equilibrium condition (294) was derived by F. London[48] by a method which is both simple and instructive. He considered a process in which superfluid mass is changed into normal fluid mass such that the total mass, momentum, and energy are conserved. Thus,

$$\delta\rho_s + \delta\rho_n = 0,$$

$$\delta(\rho_s \mathbf{v}_s) + \delta(\rho_n \mathbf{v}_n) = 0,$$

$$\delta(\tfrac{1}{2}\rho_s \mathbf{v}_s{}^2) + \delta(\tfrac{1}{2}\rho_n \mathbf{v}_n{}^2) + \rho\delta e = 0. \qquad . \qquad . \qquad (302)$$

Since $\mathbf{v}_s$ is the velocity field associated with some macroscopic quantum state, e.g., the Bose condensate, we must have $\delta\mathbf{v}_s = 0$ in the above process, i.e., except for possible boundary conditions that we can only associate with the superfluid mass. Therefore, from the second equation,

$$\delta\mathbf{v}_n = -\frac{1}{\rho_n}(\mathbf{v}_n - \mathbf{v}_s)\delta\rho_n.$$

Substitution in the energy equation (302) gives

$$\rho\delta e = \tfrac{1}{2}(\mathbf{v}_n - \mathbf{v}_s)^2\delta\rho_n = -\tfrac{1}{2}(\mathbf{v}_n - \mathbf{v}_s)^2\delta\rho_s, \qquad . \qquad (303)$$

meaning, as $\delta\rho_n = \rho\delta\alpha$, that one recovers (294).

Let us consider further the equilibrium condition (303) in the following form in situations where the normal fluid velocity vanishes:

$$n\delta g^* + \tfrac{1}{2}m\mathbf{v}_s{}^2\delta n_s = 0; \qquad . \qquad . \qquad (304)$$

here m is the mass of a helium atom, and $g^* = mg$ the chemical potential per atom, while n and n_s are, respectively, the density and superfluid density of atoms. The quantities $\mathbf{v}_s$ and n_s can be related to a wave function describing the macroscopic quantum state, $\Phi = Re^{i\phi}$, where R and ϕ are real. Hence

$$\mathbf{v}_s = \frac{\hbar}{m} \operatorname{grad} \phi$$

and
$$n_s = |\Phi|^2 = R^2. \qquad . \qquad . \qquad (305)$$

The total generalized Gibbs potential is

$$G = \int d^3\mathbf{x}(ng^* + \tfrac{1}{2}mn_s\mathbf{v}_s{}^2) = \int d^3\mathbf{x} \left(ng^* + \frac{\hbar^2}{2m} R^2 (\operatorname{grad} \phi)^2 \right).$$

To make this expression consistent quantum-mechanically, the term in the velocity should be generalized by addition of $\dfrac{\hbar^2}{2m}(\operatorname{grad} R)^2$. Furthermore, in a system near the transition point, where n_s is small, the chemical potential ng^* can be expanded in powers of n_s; and the zero-order term will be the dispensable potential of the ordinary fluid. We thus consider, in place of the above G, the function

$$\Delta G = \int d^3\mathbf{x} \left[an_s + \tfrac{1}{2}bn_s{}^2 + \frac{\hbar^2}{2m} \left(R^2 (\operatorname{grad} \phi)^2 + (\operatorname{grad} R)^2 \right) \right],$$
$$. \qquad . \qquad . \qquad (306)$$

wherein a and b are temperature-dependent coefficients. Expressing (306) in terms of the wave function Φ by means of (305), we may then regard the following variational principle as a generalization of (304):

$$\delta(\Delta G) = \delta \int d^3\mathbf{x} \left(a|\Phi|^2 + \tfrac{1}{2}b|\Phi|^4 + \frac{\hbar^2}{2m} |\operatorname{grad} \Phi|^2 \right) = 0. \quad (307)$$

It is seen to lead to the Gross–Pitaevsky[52] equation

$$\left(a + b|\Phi|^2 - \frac{\hbar^2}{2m} \operatorname{div} \operatorname{grad} \right) \Phi = 0. \quad . \qquad . \quad (308)$$

If we are dealing with charged particles, as in a superconductor, moving in a magnetic field with vector potential $\mathbf{A}$, equation (307) must be further generalized to

$$\delta \int d^3\mathbf{x} \left[a|\Phi|^2 + \tfrac{1}{2}b|\Phi|^4 + \frac{\hbar^2}{2m} \left| \left(\operatorname{grad} + \frac{ie^*}{\hbar c} \mathbf{A} \right)\Phi \right|^2 \right.$$
$$\left. + \frac{1}{8\pi} (\operatorname{curl} \mathbf{A})^2 \right], \qquad . \qquad . \qquad (309)$$

where e^* is the charge on the particles and the last term is the field energy. In a superconductor, as shown by the theory of Bardeen, Cooper and Schrieffer,[53] the supercurrent is carried by electron pairs, so that $e^* = 2e$. Consideration of independent variations of Φ and $\mathbf{A}$ in (309) leads to the well-known Ginzburg–Landau[54] equations of superconductivity, viz.,

$$\left[a + b|\Phi|^2 - \frac{\hbar^2}{2mc} \left(\mathrm{grad} + \frac{ie^*}{hc} \mathbf{A} \right)^2 \right] = 0$$

$$\frac{1}{4\pi} \, \mathrm{curl} \, \mathrm{curl} \, \mathbf{A} = - \frac{ie^*h}{2mc} \, (\Phi^* \, \mathrm{grad} \, \Phi - \Phi \, \mathrm{grad} \, \Phi^*) + \frac{e^{*2}}{mc^2} \, |\Phi|^2 \mathbf{A}.$$

$$\cdot \qquad \cdot \qquad \cdot \quad (310)$$

Vortices

The two-fluid model of He II discussed in the previous section accounts very satisfactorily for many of the hydrodynamic properties of the superfluid. However, the theory has one serious defect: it predicts that the superfluid flow will always be irrotational, i.e., curl $\mathbf{v}_s = 0$, whereas it is found that the superfluid can quite easily be brought into rotation. This difficulty is removed as soon as it is realized that, although the irrotational condition must hold throughout most of the superfluid, it is possible to have singularities in the velocity field in the form of vortex lines. The wave function in (308) would then have nodes at the positions of the vortex lines. This suggestion was first put forward by Onsager[55] and developed later by Feynman.[56] Onsager also pointed out that the superfluid circulation around a vortex line is quantized in units of h/m. This follows immediately from equation (305), relating $\mathbf{v}_s$ to the gradient of the phase of the wave function. The circulation around some contour in the superfluid is

$$\oint \mathbf{v}_s \, . \, d\mathbf{l} = \frac{\hbar}{m} \oint \mathrm{grad} \, \phi \, . \, d\mathbf{l} = n \, \frac{h}{m}, \qquad . \qquad (311)$$

where n is an integer. The last step is valid because the wave function must be single-valued, so that in going around any contour in the superfluid ϕ can change only by 2π times an integer. If the He II is contained in a volume that is simply connected, the circulation vanishes. In the presence of vortices the volume becomes multiply connected. A more accurate description of a vortex is obtained by solution of (308), which has been carried out by Gross and Pitaevsky.[52] We will discuss the motion of vortices in the superfluid near $T = 0°$ K, so that the effect of the normal fluid is negligible. The discussion also applies to vortices in ordinary fluids where viscous effects are unimportant.

The study of the motion of isolated vortices is based upon the well-known result of Helmholtz and Kirchhoff,[57] which states that the vortex moves in the fluid at the local fluid velocity, the direct effect of the velocity field of the vortex itself being disregarded. A number of problems of vortex motion can be treated directly from this law, but nevertheless it is interesting to consider vortex motion from the point of view of a variational principle.

It was shown by Kirchhoff[57] that the motion of n rectilinear vortices of strengths k_i in an unbounded incompressible fluid is given by†

$$\rho k_i \frac{dx_i}{dt} = -\frac{\partial W}{\partial y_i} \quad \text{and} \quad \rho k_i \frac{dy_i}{dt} = \frac{\partial W}{\partial x_i}, \quad (i = 1, 2 \ldots n)$$

. . . (312)

provided W is defined as

$$W = \frac{\rho}{2\pi} \sum_{i>j} k_i k_j \log r_{ij}, \quad r_{ij} = \sqrt{(x_i - x_j)^2 + (y_i - y_j)^2},$$

. . . (313)

a function of the position coordinates x_i, y_i of the vortices. The function W, sometimes referred to as Kirchhoff's stream function, is the negative of the energy of interaction of the n vortices per unit length. Thus the fluid velocity due to the two-dimensional distribution of isolated vortices can be derived from the stream function

$$\psi(r) = \frac{1}{2\pi} \sum_i k_i \log |\mathbf{r} - \mathbf{r}_i|$$

by differentiation; we obtain

$$v_x(\mathbf{r}) = -\frac{\partial \psi}{\partial y} \quad \text{and} \quad v_y(\mathbf{r}) = \frac{\partial \psi}{\partial x}.$$

The total kinetic energy per unit length is

$$\frac{\rho}{2} \int (\operatorname{grad} \psi)^2 dx dy = -\frac{\rho}{2} \int \psi \operatorname{div} \operatorname{grad} \psi \, dx dy$$

$$= -\frac{\rho}{2\pi} \sum_{i>j} k_i k_j \log r_{ij} = -W,$$

where the self-energies of the vortices have been omitted. The above results (312) and (313) have been extended by Lin[58] to the case of rectilinear vortices in a bounded fluid.

† In this section we drop the summation convention on repeated indices.

Equation (312) is exactly analogous to Hamilton's equation in classical mechanics, and it follows immediately from the variational principle

$$\delta \int_{t_0}^{t_1} dt \, (W - \sum_i \rho k_i x_i \dot{y}_i) = 0. \qquad . \qquad . \qquad (314)$$

Thus $(\rho k_i)^{\frac{1}{2}} y_i$ can be regarded as a coordinate, and its conjugate momentum is $p_i = (\rho k_i)^{\frac{1}{2}} x_i$. With this formulation the variables may very easily be quantized. The Heisenberg uncertainty relation between a coordinate and its conjugate momentum now gives for the uncertainty in the position of a vortex line of unit length,

$$\Delta x_i \Delta y_i = \frac{\hbar}{\rho k_i}.$$

For a vortex of length L this must be replaced by

$$\Delta x_i \Delta y_i = \frac{\hbar}{\rho k_i L}. \qquad . \qquad . \qquad (315)$$

For vortices in superfluid He II, where $k_i = h/m$, this uncertainty product is $m/2\pi\rho L \approx 10^{-24} L^{-1}$, which is an exceedingly small quantity. In general, a much greater uncertainty in the position of a vortex arises from the atomic nature of the fluid.

If we consider the case of two identical vortices of length L and circulation k, the "Hamiltonian" is a function of

$$\frac{(p_1 - p_2)^2}{\rho k L} + \rho k L (y_1 - y_2)^2.$$

This has exactly the form of the Hamiltonian for an harmonic oscillator, and thus

$$(x_1 - x_2)^2 + (y_1 - y_2)^2 = \frac{2\hbar(n + \frac{1}{2})}{\rho k L}, \qquad . \qquad (316)$$

n being an integer. Thus the distance separating two vortices is quantized, but the length $(\hbar/\rho k L)^{\frac{1}{2}}$ here involved is too small to accord much practical significance to this fact.

§ 14

The Significance of Variational Principles in Natural Philosophy

It has hitherto been the main theme of this discourse to treat the mathematical development of variational principles in its historical and methodological perspectives. Attention was moreover directed to the related domain of theoretical physics because of its vital importance for modern theory. But we have deliberately refrained from encroaching upon the intriguing though less exact general and philosophic implications which some speculatively minded scientists have ascribed to action principles.† Considerations devoted to the cognitive value of such postulates are however indispensable to a comprehensive discussion of the subject and will be elaborated here.

Arguments involving the principle of least action have excited the imagination of physicists for diverse reasons. Above all, its comprehensiveness has appealed, in various degrees, to prominent investigators, since a wide range of phenomena can be encompassed by laws differing in detail yet structurally identical. It seems inevitable that some theorists would elevate these laws to the status of a single, universal canon, and regard the individual theorems as mere instances thereof. It further constitutes an essential characteristic of action principles that they describe the change of a system in such a manner as to include its states during a definite time interval, instead of determining the changes which take place in an infinitesimal element of time, as do most differential equations of physics. On this account, variational conditions are often termed

† In this section, too, we extend the phrase "principle of least action" to Hamilton's principle and other variational theorems.

162

"integral" principles as opposed to the usual "differential"† principles. By enforcing seemingly logical conclusions upon arguments of this type, it has been claimed that the motion of the system during the whole of the time interval is predetermined at the beginning, and thus teleological reflections have intruded into the subject-matter. To illustrate this attitude: if a particle moves from one point to another, it must, so to speak, 'consider' all the possible paths between the two points and 'select' that which satisfies the action condition. A similar orientation presents itself in the concept of a process being accomplished with the minimum expenditure of some quantity; it will be recalled that from such ideas originated the first vague enunciation of the principle of least action by Maupertuis. In this spirit, a few distinguished scientists have adhered to a theologico-metaphysical interpretation of minimum laws, although not nearly to the same extent as formerly.

Toward the end of the last century, Helmholtz invoked, on purely scientific grounds, the principle of least action as a unifying natural law, a *leit-motif* dominating the whole of physics.[22, 35] Besides recognizing its great value in mechanics and optics, he had applied it to electrodynamics and had attempted, as previously shown, to extend its range even to thermodynamics.

> From these facts we may even now draw the conclusion that the domain of validity of the principle of least action has reached far beyond the boundaries of the mechanics of ponderable bodies. Maupertuis' high hopes for the absolute general validity of his principle appear to be approaching their fulfilment, however slender the mechanical proofs and however contradictory the metaphysical speculations which the author himself could at the time adduce in support of his new principle. Even at this stage, it can be considered as highly probable that it is the universal law pertaining to all processes in nature. . . .   In any case, the general validity of the principle of least action seems to me assured, since it may claim a higher place as a heuristic and guiding principle in our endeavour to formulate the laws governing new classes of phenomena.[22]

While Helmholtz conceded that *irreversible* thermodynamic processes were outside the immediate province of the action principle, he yet maintained that such processes were actually composed of motions of molecules and might therefore be regarded as mechanical, reversible and amenable to the principle of least action. Hence this law of physics retains its universal character.

† We are referring to any differential equation of physics, not merely to the differential principles of §9.

Among the staunchest protagonists of a metaphysical content assignable to the action principle was Planck, who, with great philosophic poise, sought first to clarify and then to extol its rank in natural science. He was particularly elated that the rich gamut of phenomena could be subsumed under what he regarded as so simple a principle,—a contemplation in harmony with his whole outlook. "As long as there exists a physical science," he meditates,

> it has as its highest and most coveted aim the solution of the problem to condense all natural phenomena which have been observed and are still to be observed into one simple principle, that allows the computation of past and more especially of future processes from present ones. It is in the nature of the case that this aim has neither been reached today, nor will it ever be completely attained. . . .
>
> Amid the more or less general laws which mark the achievements of physical science during the course of the last centuries, the principle of least action is perhaps that which, as regards form and content, may claim to come nearest to that ideal final aim of theoretical research.[36]

It may be contested that the principle of the conservation of energy vies with that of least action for this privileged position. Planck, however, pointed to the well-known fact that the energy principle provides only one equation for the changes of a system, whereas the principle of least action (in the Hamiltonian form) furnishes sufficient equations to specify fully these changes and, indeed, contains the law of the conservation of energy as one of its results. In striking contrast to his otherwise calm and balanced judgement, he dubbed the principle of least action the "most comprehensive of all physical laws which governs equally mechanics and electrodynamics." Examples of this enthusiasm abound in Planck's work.[37] Thus, he contended that the whole complex of differential equations in mechanics, electrodynamics and even thermodynamics is all included in one single law. In this statement, he certainly transgressed the bounds of experimentally verified theory, since the efforts of Helmholtz to apply the principle to thermodynamics cannot be adjudged successful. Planck elsewhere limited the scope of the action principle to reversible processes, but maintained that it reigns supreme in this field.

Whilst, in his appreciation of the unique place which the principle of least action holds in physics, Planck's views are steadfast and consistent, his advocacy of a teleological interpretation of this law is characterized by a certain measure of contradiction. In a lecture on "Religion and Natural Science" he takes a decided stand in favour of teleology and attempts to employ it as an argument for the existence of the Supreme Being.[38] Such a principle, he asserts, suggests to a person free from prejudice the presence of a rational, purposive will governing nature, for a physical system must choose that route which directs it most easily towards its objective.

> In fact, the least-action principle introduces an entirely new idea into the concept of causality: The *causa efficiens*, which operates from the present into the future and makes future situations appear as determined by earlier ones, is joined by the *causa finalis* for which, inversely, the future—namely, a definite goal—serves as the premise from which there can be deduced the development of the processes which lead to this goal.

Here Planck tries to concatenate action principles with causality and determinism—subjects ever dear to his heart—and to readmit by means of mathematical reasoning a concept of causality foreign to modern science. On this issue, he contrasts the viewpoint of physics with that of philosophy. With respect to the former, it is in vain to ask which notion of causality is the truer, and the choice between the two approaches is to be made merely on pragmatic grounds; whereas from a broader aspect, he continues, we are confronted with a formulation of causal relations that portrays a definite teleological nature.

Like trends of thought may be found scattered in Planck's popular and academic writings and addresses. An outstanding example is the extraordinary manner in which he comprehends as a variational principle Leibniz's maxim that our world is the best of all possible worlds! The strong resemblance of this comparison to the mental outlook which led to the discovery of the principle of least action is distinctly evident. But, it must be stressed, a variational principle as now conceived is a highly specialized type of theorem and does not simply represent the minimization or maximization of some quantity—be it the Lagrange function or the eternal good. Whatever our opinion concerning the objective validity of such purely theological ideas, we cannot do other than wonder at the vigour with which the founder of the greatest revolutionary theory of our age could defend so conservative a standpoint and accept the stigma of an outmoded *Weltanschauung*.

Yet elsewhere he ardently repudiated the claim that teleological factors could explain natural phenomena *via* variational principles. As before, he asserted that the principle of least action is of much avail from the practical angle, since it is independent of the choice of co-ordinates, but, in accordance with his habit of exact reasoning, he drew attention to the fallacy of investing it with teleological implications. With this guard against conferring metaphysical profundity upon a law of physics, he escaped, in the present instance at least, the danger of artificially applying specific theorems of experimental science to philosophic thought in general.

With the development of the older form of quantum theory, the persuasion that the action function had some deeper meaning gained renewed impetus. In another part of this monograph we have mentioned how the concept of action returned in quantum theory, subject to the puzzling and remarkable restriction that it could occur only in whole multiples of the elementary quantum of action. As a result of this dual appearance of the action integral in dynamics and quantum theory, the crudity of Pythagorean number mysticism seemed to have been superseded by a finer, more subtle variety. For, proclaim its adherents, in the phenomena of the material universe itself we now apprehend an entity from which multifarious propositions may be deduced if we postulate that, on the one hand, it is to be subjected to minimum conditions and, on the other, it should be allowed to assume integral values only. In Sommerfeld's words:

> Our spectral series, dominated as they are by integral quantum numbers, correspond, in a sense, to the ancient triad of the lyre, from which the Pythagoreans 2500 years ago inferred the harmony of the natural phenomena; and our quanta remind us of the role which the Pythagorean doctrine seems to have ascribed to the integers, not merely as attributes, but as the real essence of physical phenomena.[39]

However, it was again Planck who pursued this train of thought with more fervour than did any other physicist. We recall that, even before Sommerfeld and Wilson introduced the quantization of phase integrals, Planck, when dealing with the division of phase space, had connected his quantum theory with the principle of least action and intimated that the action was associated with whole multiples of h, which he designated the "elementary quantum of action." At a later period, after the relationship between the action integral and quantum theory had been made more explicit, Planck posed the question as to the more penetrating implications of this situation. "Can it be that the astonishing simplicity of this relation rests once again upon chance? It is becoming more and more difficult to believe this. On the contrary, the impression forces itself upon us with elemental power that Leibniz's† principle of least action can afford the key to a deeper understanding of the quantum of action."

† *Vide* p. 23.

Planck's arguments concerning this problem are steeped in considerable metaphysical hypothesis to which it is difficult for the critical scientific reader to give assent. Whatever value his actual contribution may possess, however, his life-long pre-occupation with the wider aspects of variational principles and his refusal to acquiesce in their traditional presentation as heuristic methods of mathematical physics have, beyond doubt, caused natural philosophers to concentrate on their less obvious features and have thus stimulated research.

Certain other contemporary scientists, too, have put forward similar speculative assumptions, although not nearly with the same zeal and persistence as Planck. In the complicated metaphysical system of Whitehead,[40] appeal is made to the court of philosophic reflection to justify the *raison d'être* of the principle of least action. It is nevertheless in agreement with his general outlook when he declares that this principle is aprioristically determined only as regards its abstract form and not its detailed character which must be obtained by experiment. In assuming this position, he overcomes the difficulty engendered by the apparent arbitrariness of the function which is to be minimized.

Proceeding from a different viewpoint, the biologist Bertalanffy[41] attempts to persuade us that the principle of least action looms in spheres other than dynamics, and quotes, in support of his assertion, Lenz's law in electricity, the principle of Le Chatelier in physical chemistry and Volterra's theory of population. No more and no less does he demand than that the action principle be generalized to embrace systems of all kinds—open or closed, organic or inorganic. Such an intention could be appreciated in a logico-formalistic scheme, but never in a "unified science," where the structural configurations symboliz-ing the individual disciplines are surely meant to correspond with empirical data. Besides, to identify the aforementioned laws with action principles amounts again to diluting the meaning of the term so freely that its essence is lost and it can be made to include almost anything.

While the academic worth of the contributions by Whitehead and Bertalanffy is in no way disputed, our reference to them must needs be cursory, as variational principles play but an incidental part in their philosophy. We have now reached the stage where we are able to undertake a critical methodological examination of the anthropomorphic and metaphysical super-structure which even a Planck could erect upon these mathe-matized theories, and where we can turn to an epistemological evaluation of variational principles, which is so frequently prejudiced by premature philosophical commitments. As a preliminary step, however, it is necessary to scrutinize the Pythagorean and neo-Platonic notions of simplicity, order and regularity prevailing in the world of physical objects.

From the days of old, man has fostered the idea that rational order lies at the base of all things. We have already recounted in brief how this and similar thoughts were promoted and co-ordinated by the Pythagorean school and by Plato, who organized them into a more systematic philosophic edifice. In classical, Greek cosmology, the propensities of order and harmony were imagined to inhere in the cosmos itself, quite independently of the human mind. The emancipation of man's intellect from total reliance on the regularity and perfection of the material universe was perhaps the most revolutionary achievement of the post-scholastic era, when Roger Bacon paved the way for humanity's self-liberation. The temptation to search for patterns of regularity in nature was balanced by the equally powerful inclination to attribute to human reason a predilection for simplicity and mathematical definiteness. But pioneers of science like Copernicus, Kepler and Galileo realized—and here they differed most essentially from the Greeks—that the "mysteries" of nature could be illumined only with the assistance of accurate observations and experiments. It was just this twofold approach that was responsible for their epochal successes, which became milestones in the history of science. These remarks are, of course, not meant to be anything but the barest gist of the complicated philosophic process which was taking place at the dawn of the modern age.

The conviction that a mechanical view could provide an exhaustive explanation of natural phenomena was of necessity accompanied by the belief that the simple order of scientific laws was not only a prerequisite for the systematic classification of empirical data, but reflected moreover the fundamental order existing within the realm of objective reality. Any historian of science will bear out that the advance of physics during the two centuries following the publication of Newton's *Principia* was made possible largely by this presupposition. A particularly outstanding illustration is presented in Maxwell's equations of the electromagnetic field, that, in a sense, constitute the zenith of classical physics. The aptitude with which he combines the experimental results of Faraday and an intuitive leaning toward symmetry may well serve as a model of the scientific method.

With a boldness rare among scientists, the Austrian physicist and natural philosopher Mach (1838–1916) began the crusade against this universally accepted approach of classical mechanics and, in his eagerness to purge physics of its scholastic, i.e., metaphysical relics, was carried right to the other extreme. According to his phenomenalistic theory of knowledge, physical "reality" can never be causally explained; the scientific theory can do no more than describe systematically the simple sense data and the relations between them. Mach's criterion for a proposition to have scientific, that is, intelligible meaning, is that it must be possible to formulate it as a statement about sense impressions. Philipp Frank, the foremost interpreter of the positivistic philosophy of Mach, stresses that it is necessary only *in principle* for such statements to be reducible to statements about sense perceptions and draws the analogy with a theorem in the theory of functions, which could, *in principle*, be expressed as a theorem about the integers. The physicist—Mach criticized—can merely register sensory phenomena and confirm his predictions by observation. To pass from sense experience to a hypothetical metaphysical substratum, as for example matter, is therefore logically impossible and amounts to a superfluous detour. The question as to the "objective" content of a theory is not only relegated to the background, but is declared meaningless: the constant elements of physical knowledge consist solely in a system of relations or equations.

It is easy to understand that in the doctrines of a philosopher displaying such an attitude there is no place for the concepts of simplicity, regularity and perfection inherent in nature itself. As against this, we should strive for the utmost simplicity in our analysis of nature, because in phenomenalistic physics, in contrast to mechanistic physics, a theory has no further function than to yield an economical and convenient *modus* of interrelating sense data. "The basis," writes Mach, "of all my investigations into the logical foundation of physics as well as into the physiology of perceptions has been one and the same opinion: that all metaphysical propositions must be eliminated, because they are idle and disturbing to the economical design of science."

While the criticisms levelled by modern philosophers of science against Mach cannot be dismissed with impunity, and are, in part at least, of decisive significance, even a summary examination of these would lead us into a region much too large and remote from our subject to be discussed adequately in a single chapter. It cannot easily be ignored, however, that Mach discriminated too drastically between exact science and "transcendence" or metaphysics, as for instance, in his rejection of the atomic theory; moreover, his positivistic views failed to do justice to the universal pervasiveness of mathematics in physics. Cassirer has shown that, for all its sobering influences, *phenomenological* physics has its roots not in the logic of physical science, but in anthropomorphic rudiments and in psychologism— emotional foundations of no more scientific cogency than speculative philosophy of nature.

Aprioristic and materialistic science alike were also denounced by the French mathematician, physicist and philosopher Poincaré (1853–1912), whose "conventionalism" is less unattractive to the mathematically minded investigator than Mach's phenomenalism. A scientific principle was understood by Poincaré to be simply a convention or a definition in disguise, the result of a free process of human thought, with the aid of which we interpret and systematize the empirical facts. Since the validity of a physical principle cannot be ascertained on the basis of experiment and observation alone, it is neither true nor false; it is merely a convenient generalization. But principles are emphatically not *arbitrary* artifacts, because, from the infinite number of logically possible conceptual patterns, we select only those which are of avail in classifying and integrating the data of experience. Thus, axioms of physics, at which we arrive when seeking for a common denominator of a great many laws, are the "quintessence of innumerable observations"; to endow them with the property of universality is a free creation of man's mind.

With exemplary clarity and precision did Poincaré define his conception of the ultimate goal of physical knowledge.[42]

> Science is, in other words, a system of relations. It is only in relations that we should attempt to find objectivity; it would be futile to search for it in the things themselves instead of in their relations to one another. The assertion that science can have no objective value because it provides us only with knowledge of the relations would be wrong, for it is just these relations which are to be regarded as objective.

The confidence that there is something enduring in a scientific theory was thereby not set aside,—it was only transferred from the plane of "absolute reality" to that of structural relations. Tendencies of a similar kind can be discerned in *Gestalt* psychology, in organismic biology and in Eddington's philosophy of physics, where we know "not things, but structures of relations between unknown things."

Poincaré's general agnostic outlook culminated in his profound criticism for which the notion of simplicity had been made the occasion. It was a custom of old standing, he polemizes, that the simplicity of a natural law was accepted as a criterion for its exactness. The belief in simplicity is, according to him, a prerequisite for the construction of a science, since such a belief is implicitly assumed in all generalization. For instance, whenever a certain number of observations are subsumed under a simple law, it is taken for granted that this law is valid for all cases, although infinitely many complicated laws could harmonize with the observed facts. "To sum up, in most cases every law is held to be simple until the contrary is proved."

On the other hand, it is not always evident at first sight whether a law is simple or complicated. Poincaré quotes the disturbed motion of the planets, which is extremely involved and yet governed by the most simple gravitational law of Newton, whereas Boyle's law is an apparently simple relation behind which is concealed a vast complexity of detail. The simplicity of Boyle's law is thus due to the interplay of large numbers, i.e., each individual fact is intricate, but the mean of all the effects is simple. Furthermore, a relation of direct proportionality sometimes occurs for no other reason than that one of the variables concerned suffers a small increment, and not because of any simplicity intrinsic in the natural process. Such a proportionality would, nevertheless, be only approximate. It seems, he asserts, as if the constant advance of science, which increases the accuracy of our knowledge and continually opens up new vistas, tends to render the laws of nature more complex. Poincaré therefore concludes that the doctrine, current fifty years before, of a simple law being more probable than a complicated law, had been forsaken; yet physicists still had to act as though they were ruled by it. With the unifying discoveries in relativity and quantum mechanics, however, the situation has radically changed, and we find a great wealth of physical phenomena embraced by mathematical theories containing comparatively few postulates. But unlike the classical mechanics of Galileo and Newton, such theories are of a highly abstract nature and the simplicity is far removed from the plane of human experience.

These comments on the concept of simplicity should not be taken to imply that any measure of concordance exists among contemporary scientific thinkers in regard to this issue. Whitehead, otherwise antipodal to the French neo-positivists, on this question entertains views not dissimilar to those of Poincaré. He contrasts the complexity of the facts themselves with the goal of science, which is to search for their simplest explanation and which motivates us to perceive simplicity where it does not prevail—"Seek simplicity and distrust it." Proceeding from basic conceptions much closer to those of positivism, Bridgman, while he conceded that a large part of physics consisted in the contraction of complicated data to simpler theories, deems the evidence to be overwhelmingly against simplicity immanent in nature. For, simplicity can mean only simplicity in terms of our mode of thought, and to assume that nature behaves in a simple manner is to assume that it acts in accordance with our thinking processes, a highly unlikely occurrence owing to the very indirect and inexact relation which, he maintains, our immediate concepts bear to nature itself. He explains the observed simplicities by the limited range and therefore restricted type of phenomena accessible to our senses.

In the face of these attempts to give rational arguments for the widely held tenet that simplicity is not a concomitant property, but an essential attribute of either the human intellect or of objective nature, we must ask the question: have we any scientific and philosophic justification for the conviction that there is simplicity at all? Unless we are prepared to clarify this central problem, a way out of the impasse seems impossible. The whole discussion has been prejudiced through lack of a precise definition concerning the signification of the word "simplicity." True, the progressive amassing of empirical information is accompanied by the diminution in the number of necessary hypotheses, and more phenomena are continuously being encompassed by fewer theories; as Leibniz put it: "The sciences concentrate as they expand." This undeniable fact has time and again been invoked in support of the simplicity postulate. But Leibniz's maxim is ambiguous if not applied with due caution; it does not hold, for example, in the biological sciences, and in many of the physical sciences the reduction to the elementary hypotheses can be brought about only in principle, because of the intricacy of the mathematics. This limitation aside, the argument for simplicity is restricted to mathematized theories of natural science, where we deal with measurable and exact entities. Even here, one can attain no more than a vague degree of conciseness regarding the simplicity of a theory. To illustrate this point, let us cite Born's example: which law of gravitation is simpler, that of Newton or that of Einstein? Whereas physicists of the old school may have decided in favour of Newton, the modern scientist would probably reply that, notwithstanding its abstractness and complicated mathematics, Einstein's law was the simpler, because the additional concept of force did not have to be introduced. In general, recent theories are certainly not simple in the sense that their application to everyday phenomena can be carried out easily—on the contrary, the relation between the fundamental constituents of the theoretical system and these phenomena is most indirect.

Thus the simplicity of a mathematized theory can be assessed only by means of a relative and purely arbitrary definition. And while we must concur with Weyl when he ranks the simplicity postulate of foremost import for the epistemology of physical science, we should beware of inadvertently broadening the concept beyond its definition into a realm where it forgoes its purely formal and dispassionate meaning. After all is said, we can readily appreciate how delicate and subtle the problem of simplicity becomes when studied as a physical concept. However, if we succumb to the temptation of adopting the term "simplicity" as it is most generally used—on a par with words like elegance, harmony and beauty—we are immediately guilty of a "gross relapse into a form of anthropomorphism" (Cassirer). Whether psychological, aesthetic or philosophical factors form the elements of such an anthropomorphic predisposition is without relevance for the ultimate effect; the approach will in any case lead eventually to scholastic, dogmatic metaphysics. Even a scientist of Weyl's calibre refers to "profound harmony" and "true inner simplicity," phrases which are dangerous because of their metaphysical insinuations.

It is similarly a traditional misapprehension to construe an accord between the postulate of simplicity and 'least' principles in nature. To declare in earnest that there is some deeper meaning attached to the idea of nature's acting in such a way that it makes some quantity a minimum, is perhaps a relic of the times when the universe was believed to be driven by a supra-human being. Obsessed with the urge to achieve "cognition of real nature," i.e., to gain an "explanation" of material pheno-mena, physicists have assigned to a particle the knowledge which enables it to take the most convenient path of all those which would direct it to its destination. The Aristotelian final cause is dragged into the subject, and the particle is presented to us as what Poincaré calls "a living and free entity." Aristotle's dictum that nature does nothing in vain may have been at the root of Maupertuis' teleological principle, which initiated the study of variational principles in nature, but to the twentieth-century physicist the notion of such a mysterious purposive agency would only be a regress into sheer obscurantism. History knows of many instances where a science had its beginnings in a mythical and magical symbolism which gradually shed its mystical features as our rational knowledge increased and our conceptions became clearer—witness the transition from astrology to astronomy and from alchemy to chemistry. With Cassirer we may say that, "It was a false and erroneous form of symbolic thought that first paved the way to a new and true symbolism, the symbolism of modern science." At the end of the 18th century, Kant could, in opposition to Spinoza, still defend the plea for teleology, though as a strictly regulative and heuristic principle, not as an objective characteristic inherent in nature. Whatever value such a restricted interpretation of the idea of purpose may have for the speculative systems of certain philosophic schools, Kant's critical reasoning, with all its concessions to scientific thought emancipated from metaphysical associations, can today not be accepted by mathematical physics.

The belief in a purposive power functioning throughout the universe, antiquated and naive as this faith may appear, is the inevitable consequence of the opinion that minimum principles with their distinctive properties are signposts towards a deeper understanding of nature and not simply alternative formulations of differential equations in mechanics. There is moreover no way of telling beforehand *which* quantity is to be made a minimum. Maupertuis, using imagination and speculative reasoning, admittedly arrived at a definition of action that could be considered an inexact version of the correct definition, but, as already emphasized, his conjecture was unworkable and, in any case, he failed to anticipate the energy restriction, which was indispensable to the Euler-Lagrange form of the principle of least action. Euler similarly claimed that he had derived his action function partly intuitively, partly from inductive inference; be it as it may, some of the previously observed laws of mechanics were incorporated through the energy condition. And when we turn to the variational principles which the modern physicist values most, such as that of Hamilton or those of electromagnetism, there is no question of apriorism in the determination of the relevant function. Now, assuming a teleological force to be at the root of minimum principles, one would surely be able to recognize by some other criterion why nature has chosen this particular and no other quantity to be kept at a minimum. This argument, so frequently put forward, may not by itself be a sufficient refutation of purposive explanations in physics, but even Bavink, who is not averse to resorting to metaphysics in the analysis of biological phenomena, refused on these grounds to interpret minimum principles teleologically.

Much as these reflections may serve to discredit the rôle of purpose in 'least' principles, they are indeed supererogatory, since the so-called minimum principles are not minimum principles at all, but variational principles! They merely affirm that the variation of the integral under consideration—if the dependent variables are given a slight change, subject to certain boundary conditions—is zero, or more strictly, is an infinitesimal quantity of the second order. This is, as we have repeatedly stressed, not tantamount to maintaining that, of all the possible ways in which a system can change from one state to another, it actually follows that way for which the integral is a minimum. When the minimum condition holds, the variational condition certainly must hold, but *the converse is not true.* In the case of the principle of least action (Lagrange's form) or of Hamilton's principle, a real minimum is obtained provided the path of the system is short enough. Otherwise neither a maximum nor a minimum may occur. In the simplest case, namely that of a particle free from forces, the action is always a minimum, the path being a straight line. As an instance of where the action is not a minimum, one need merely cite the case of a particle travelling in an ellipse under the influence of a Coulomb force. The particle may move between two points on the ellipse in either of two paths; the energy is the same in both cases, but both paths cannot have the *least* possible action.

Hence the teleological approach in exact science can no longer be a controversial issue; it is not only contrary to the whole orientation of theoretical physics, but presupposes that the variational principles themselves have mathematical character- istics which they *de facto* do not possess. It would be almost absurd to imagine a system guided by a principle of purpose in such a manner that sometimes, not always, the action is a minimum.

Untenable as the arguments in support of teleological ideas underlying variational principles have proved, it still remains for us to inquire into the so-called all-embracing nature of these laws. Now it cannot be contested that variational principles do enable us to condense several mathematical equations describing physical phenomena into one compact theorem; furthermore, we behold here one type of principle which governs the laws of a large part of physics. This fact is, however, by no means extraordinary, because these branches of physics are all characterized by differential equations, and it is not at all an exceptional property of a set of differential equations to be expressible as a variational principle. As a matter of fact, most sets of (mathematically) simple differential equations can be so formulated.   Conversely, any variational principle is equivalent to a set of differential equations.   It follows immediately that once the laws of a physical theory are expressed as differential equations, the possibility of their reduction to a variational principle is evident from purely mathematical reasoning and does not depend on certain attributes intrinsic in the theory. This possibility is of pragmatic value, for, other factors aside, we have seen that it allows the equations to be written in such a way as to be independent of the co-ordinate system.   In addition, the transformation of the equations to the Hamiltonian form is vitally important for quantum mechanics.   But it would be unwarranted to go beyond the heuristic and generalizing function of variational principles in mathematical physics.

The equivalence of a variational principle to its corresponding set of Euler equations is also sometimes regarded as an argument against interpreting differential equations as indicative of cause and integral principles as attesting purpose, since the transformation from the integral to the differential form would then amount, as Margenau puts it, to transmuting a purpose into a cause.   While this refutation may strike one as not altogether convincing, we must none the less concur with his view that such apparently profound antitheses are meaningless unless understood strictly analytically.   Whether the differential or integral formulation is employed depends on convenience alone.

On account of these critical observations, it is not too much to say that, if certain scientists are constrained to indulge in metaphysical daydreams about the action principle, their meditations, though perhaps interesting, are nevertheless scientifically unintelligible.

The assertion made above, that a physical theory can unconditionally be represented by a variational principle, requires some qualification in the case of electrodynamics, where, as we pointed out at the time, the same Lagrangian serves to derive the equations of the field and of the charged bodies, a conclusion in no way mathematically self-evident. This circumstance is connected with the fact that the equations of the combined field-body system can be written in Hamiltonian form, which in turn is necessary for the passage to quantum mechanics.   Actually, however, classical theories are mere approximations or analogies to the corresponding quantum-mechanical theories, and it would therefore be expected that a classical theory could be embodied in quantum mechanics. Hence one might have anticipated that the equations of classical electrodynamics are expressible as a variational principle. But this possibility does not suggest any deeper, and perforce transcendent, significance of such principles, as the equations of quantum mechanics are *not* equivalent to a variational principle —requiring a certain integral to be stationary—as treated in this book.

Generally speaking, and as already mentioned, stationary principles do not occupy in quantum mechanics the prominent place they hold in classical mechanics. The principle of least action, according to Weyl the pinnacle of classical mechanics, had within fifty years lost its pre-eminence, and Larmor's optimistic prophecy that it would outlive all the other laws of physics had been shattered. Stationary principles of a nature entirely different from the classical principle of least action appear in quantum mechanics in one association: if we are using the wave-mechanical formalism, the equations concerned are linear differential equations which can be stated as variational principles. Yet here the function whose integral is to be stationary is usually of a rather intricate and artificial character. What physicist would prefer, as the fundamental laws of wave mechanics, variational principles like, for example, that of Darwin, to the Schrödinger or Dirac differential equations? Quantum mechanics in this respect differs markedly from the older form of quantum theory, where the quantization of the action seemed to present an attractive field for idealistic pretensions. The integer restrictions have been replaced by other and more comprehensive laws not implicating any such Pythagorean concepts; the high hopes of those who displayed a predilection for metaphysics have once more failed.

Although the principle of least action does not play a great part in quantum mechanics, we have nevertheless explained how the fundamental laws of the subject, and therefore of theoretical physics, can be formulated by means of a single principle involving the action function. Whether or not this principle, in the form accorded it by Feynman or Schwinger, has some deeper significance not yet apparent, is a question which perhaps no physicist would yet venture to answer. The quantum theory of fields has now reached an impasse, from which we shall probably extricate ourselves only by introducing some fundamentally new idea in the formulation of the basic laws. The envisaged re-formulation may not necessitate an essential modification to the structure of quantum mechanics, or it may involve a change no less radical and far-reaching than that from classical to quantum mechanics. It may possibly be of such broad scope as to enable us to determine the masses and other properties of the elementary particles. We need hardly add that the direction in which we shall have to look in order to find the new formulation is completely uncertain; its discovery will undoubtedly require the genius of an Einstein or Planck. However, the Feynman and Schwinger principles are the most succinct formulations of quantum mechanics, and the Feynman principle in particular appears to contain many latent potentialities. In our view, it will not be at all surprising if these principles prove to be of some avail in reaching a solution to the problems which at present beset fundamental physics. We should emphasize again, none the less, that this would not be a manifestation of the efficacy of the action principle, but of a principle very different even in the axiomatic concepts which it employs.

There is one aspect of variational principles which need hardly be emphasized: during this whole analysis of their scope, it is only formally that a physical theory can be enunciated as a variational principle. In other words, as Born reminds us, the physical meaning of each symbol contained in the principle has to be specified before the theory can be applied in practice. This is, of course, a truism pertaining to every mathematized physical theory; still, it should be remembered as a precaution against reading into variational principles more than is contained in them.

A final criticism of variational principles had to be preceded by such exact and methodical reasoning in order that the ground might be prepared for an epistemological evaluation. First of all, are we justified in speaking loosely about variational 'principles'? If we adopt Schlick's viewpoint that knowledge can give us no more than description by natural laws, then there cannot be any question of misuse of the term "principle." But, should we tentatively identify ourselves with Cassirer in regarding principles as the fixed points of the compass necessary for our guidance among phenomena, or as maxims for combining individual facts into coherent and consistent wholes, then we are at variance with him when he includes the principle of least action in this category. The action function can be fully and satisfactorily defined in terms of the other constructs and laws of dynamics, and it is thus rather an invaluable mathematical aid than a means of interpretation. It would amount to transgressing objective evidence to consider the principle of least action a peremptory postulate of nature, or even of some branch of physical science—to consider it on a par with, say, the principle of the conservation of energy or the Schrödinger equation in wave mechanics. Denuded of all their extra-scientific encumbrances, variational principles evince greater propinquity to derived mathematico-physical theorems than to fundamental laws.

Behind all the tendencies at interpreting the principle of least action in a somewhat spiritual manner lies the desire to secure an explanation of natural phenomena. All that such an explanation could achieve, however, was to relate observed facts to other data of general experience, data with which man was already familiar. On the one hand, anthropomorphic fetishes were in vogue, especially up to the 19th century; on the other, scientists of the previous century favoured mechanical models, literal images of the external world. The developments in theoretical physics, above all in quantum theory, have finally persuaded scientists of the fallacies innate in the mechanistic view and have thus caused its relinquishment. But it seems that positivists like Duhem have, in their reaction against this approach, gone too far by overstating the symbolic character of all scientific knowledge. Pressing the natural philosophy of Mach and Poincaré to its logical conclusion, Duhem uncompromisingly denied any connexion between the symbolism of a theory, which he regarded as self-contained, and the objective reality outside us. Although it is incontrovertible that all exact theories are symbolic in structure, his outlook is so abstract, conceptualized and remote as to be almost barren for the physicist.

In his intensive study of the epistemological foundations of science, Cassirer endeavoured to restore the balance between a purely mechanistic physics and an unadulterated positivism. He, too, consistently stressed the symbolic nature of all knowledge, scientific or otherwise, yet he never countenanced the positivist extreme of denying any objective reality behind the aggregate of observations. "We cannot anticipate the facts, but we can make provision for the intellectual interpretation of the facts through the power of symbolic thought." The task of theoretical physics would therefore consist in gradually evolving the symbolic method to correspond more and more closely with the facts, and not in fathoming the depths of nature through futile search for a mysterious absolute thing-in-itself.

This critical and moderate standpoint is, in large measure, shared by prominent modern physicists like Heisenberg and Schrödinger, who, in their recent publications, have not been impervious to epistemological concerns. When Heisenberg, for instance, replaces "interpretation of nature" by "description of nature," or when Schrödinger chooses to call a theory an "adequate" model rather than a "true" one, they certainly escape the ontological pitfalls into which many a great scientist has fallen.

Planck was one of the few distinguished physicists of our day who did seek for a metaphysical explanation of natural phenomena and thought it possible for mankind to progress toward this goal, to know the *real* world. A genuine philosophic interest, inextricably pervading his scientific work, preconditioned his whole mode of reasoning and his reflections on science in general. It was inevitable that such a thinker would be prone to exalt *some* physical theorems to the height of a universal philosophic concept. Whenever scientific laws, derived from empirical data and applicable to measurable quantities, are conceptualized, they will be shorn of all their restrictions and provisos so that their meaning must be distorted. The principle of least action above all other propositions attracted, for the reasons indicated, his imagination, and was therefore singled out for this privileged treatment. How else can we account for the indubitable fact that Planck, despite his admonition that all anthropomorphism should be eliminated from exact science, himself succumbed to the very error he denounced?

Perhaps subsequent scientist-philosophers may possess the genius to decide in favour of the categorical positivist, the metaphysical realist. . . or neither. Meantime we feel that the only way open to us is to steer a middle course between Scylla and Charybdis—between Mach and Planck. We forbear from surrendering to the prejudice that metaphysical ideas *per se*, in their mundane sense, jeopardize the advance of exact science. Rather let us follow Born's credo:

> Another objection was raised against my use of the expression 'metaphysical' because of its association with speculative systems of philosophy. I need hardly say that I do not like this kind of metaphysics, which pretends that there is a definite goal to be reached and often claims to have reached it. . . . Metaphysical systematization means formalization and petrification. Yet there are metaphysical problems, which cannot be disposed of by declaring them meaningless, or by calling them with other names, like epistemology. For . . . they are 'beyond physics' indeed and demand an act of faith. We have to accept this fact to be honest.

References

1. PLATO, *Timaeus*, 29E–30, 33B, transl. F. M. Cornford. (See general reference CORNFORD 1948.)
2. ARISTOTLE, *De caelo*, II, 4, 287a, transl. J. L. Stocks.
3. HERO, *Catoptrics*, 1–5, transl. Schmidt. (See general reference COHEN and DRABKIN.)
4. GALILEO, "Il Saggiatore," *Opere*, tomo II (1744).
5. NEWTON, *Principia*, III, transl. A. Motte.
6. F. BACON, *Novum Organum*, XLI, XLV, XLIX. (See general reference COMMINS and LINSCOTT.)
7. FERMAT, *Oeuvres* (1679). (See general reference MAGIE.)
8. HUYGENS, *Treatise on Light*, transl. S. P. Thompson, pp. 42–45 (1922).
9. MAUPERTUIS, "Accord de différentes lois de la nature," *Oeuvres*, tome IV (1768).
   "Recherche des lois du mouvement," op. cit., tome IV.
10. MAUPERTUIS, "Essai de cosmologie," op. cit., tome I.
11. EULER, *Methodus inveniendi lineas curvas maximi minimive proprietate gaudentes, additamentum II, Opera omnia*, series I, vol. 24, ed. C. Carathéodory, pp. LII–LV, 298–308 (1952).
12. MAUPERTUIS, "Loi du repos," op. cit., tome IV.
13. LAGRANGE, "Application de la méthode exposée dans le mémoire précédent à la solution de différents problèmes de dynamique," (*Miscellanea Taurinensia*, t. II (1760–1761)), *Oeuvres*, tome I (1867).
14. LAGRANGE, *Mécanique analytique*, seconde partie, op. cit., tome XI.
15. HAMILTON, W. R., "Second Essay on a General Method in Dynamics," *Phil. Trans. Roy. Soc.* (1835), 95.
16. HAMILTON, W. R., "On a General Method in Dynamics," ibid. (1834), 247.
17. JACOBI, C. G. J., "Über die Reduction der Integration der partiellen Differentialgleichungen erster Ordnung zwischen irgend einer Zahl Variabeln auf die Integration eines einzigen Systemes gewöhnlicher Differentialgleichungen," (*Crelle's Journal*, **17** (1837), 97) *Gesammelte Werke*, Band IV (1886).
18. HAMILTON, W. R., "Theory of Systems of Rays," *Trans. Royal Irish Academy*, **15** (1828), 69.
   "Supplement to an Essay on the Theory of Systems of Rays," ibid., **16** (1830), 1.
   "Second Supplement to an Essay on the Theory of Systems of Rays," ibid., **16** (1831), 193.

182 *Variational Principles in Dynamics and Quantum Theory*

"Third Supplement to an Essay on the Theory of Systems of Rays," ibid., **17** (1837), 1.

"On a general Method of expressing the Paths of Light, and of the Planets, by the Coefficients of a Characteristic Function," *Dublin University Review* (1833), 795.

See Hamilton, *Mathematical Papers*, vol. I (1931).

19. SCHRÖDINGER, E., *Ann. d. Phys.*, **79** (1926), 361, 489; **81** (1926), 109.
20. DIRAC, FOCK and PODOLSKY, *Physik. Z. Sowjetunion*, **2** (1932), 468.
21. FERMI, E., *Rev. Modern Phys.*, **4** (1932), 87.
22. HELMHOLTZ, "Über die physikalische Bedeutung des Princips der kleinsten Wirkung," (*Crelle's Journal*, **100** (1886), 137; 216) *Wissenschaftliche Abhandlungen*, Band III (1895).
23. BOHR, N., *Phil. Mag.*, **26** (1913), 1.
24. SOMMERFELD, A., *Sitzungsber. d. K. Bay. Akad.* (1915), 425.
25. SOMMERFELD, A., *Ann. d. Phys.*, **51** (1916), 1.
26. WILSON, W., *Phil. Mag.*, **29** (1915), 795.
27. EPSTEIN, P. S., *Ann. d. Phys.*, **50** (1916), 489.
28. SCHWARZSCHILD, K., *Sitzungsber. d. K. Preuss. Akad.* (1916), 548.
29. EPSTEIN, P. S., *Ann. d. Phys.*, **51** (1916), 168.
30. EHRENFEST, P., *Ann. d. Phys.*, **51** (1916), 327.
31. BURGERS, J. M., *Ann. d. Phys.*, **52** (1917), 195. Diss. Leyden (1919), 242.
32. BROGLIE, L. DE, *Phil. Mag.*, **47** (1924), 446.
33. FEYNMAN, R. P., *Rev. Modern Phys.*, **20** (1948), 367.
34. SCHWINGER, J., *Phys. Rev.*, **82** (1951), 914.
35. HELMHOLTZ, "Zur Geschichte des Princips der kleinsten Action," (*Sitzungsber. d. Akad. d. Wiss. z. Berlin* (1887), 225, op. cit., Band III).
36. PLANCK, M., "Das Prinzip der kleinsten Wirkung," (from *Kultur der Gegenwart* (1915)). (See general reference PLANCK Stuttgart 1949.)
37. *Max Planck in seinen Akademie-Ansprachen*, Deutsche Akad. d. Wiss. (Berlin, 1948).
38. PLANCK, M., "Religion and Natural Science," lecture delivered in May, 1937. (See general reference PLANCK New York 1949.)
39. SOMMERFELD, A., *Die Bedeutung der Röntgenstrahlen für die heutige Physik*, p. 11 (Munich, 1925); cited in general reference WEYL 1949.
40. WHITEHEAD, A. N., *Science and the Modern World*, p. 133 (Cambridge, 1927).
41. BERTALANFFY, L. VON, *Problems of Life*, p. 201 (London, 1952).
42. POINCARÉ, H., *La Valeur de la science*, chap. xi, p. 6 (Paris, 1905).
43. HERIVEL, J. W., *Proc. Camb. Phil. Soc.*, **51** (1955), 344.
44. LIN, C. C., *Proceedings of the International School of Physics, Varenna* (New York, 1963).
45. ECKART, C., *Phys. Rev.*, **54** (1938), 920.
46. LAMB, H., *Hydrodynamics* (Cambridge, 1952).

47. LANDAU, L. D., see Landau and Lifshitz, *Statistical Physics* (New York, 1958) and Landau and Lifshitz, *Fluid Mechanics* (New York, 1959).
48. LONDON, F., *Superfluids*, vol. II (New York, 1954).
49. TISZA, L., *Phys. Rev.*, **72** (1947), 838.
50. ZILSEL, P. R., *Phys. Rev.*, **79** (1950), 309; **92** (1953), 1106.
51. KHALATNIKOV, I. M., *Introduction to the Theory of Superfluidity* (New York, 1965).
52. GROSS, E. P., *Annals of Physics*, **4** (1958), 57; PITAEVSKY, L. P., *Zh. Eksperim. i Teor. Fiz.* (U.S.S.R.), **35** (1958), 408 [Translation *Soviet Phys. J.E.T.P.*, **8** (1958), 282.]
53. BARDEEN, J., L. N. COOPER and J. R. SCHRIEFFER, *Phys. Rev.*, **108** (1957), 1175.
54. GINZBURG, V. L. and L. D. LANDAU, *Zh. Eksperim. i Teor. Fiz.* (U.S.S.R.), **20** (1950), 1064.
55. ONSAGER, L., *Nuovo Cimento*, **6** Supp. (1949), 249.
56. FEYNMAN, R. P., *Progress in Low Temperature Physics*, vol. I (Amsterdam, 1955).
57. See reference 46.
58. LIN, C. C., *University of Toronto Studies*, Applied Mathematics Series **5** (Toronto, 1943).

General References

ARISTOTLE, *Works*, ed. Ross, vol. II (Oxford, 1930).
BAVINK, B., *Ergebnisse und Probleme der Naturwissenschaften* (Zürich, 1948).
BORN, M., *Experiment and Theory in Physics* (Cambridge, 1943).
BORN, M., *Natural Philosophy of Cause and Chance* (Oxford, 1949).
BRIDGMAN, P. W., *The Logic of Modern Physics* (New York, 1948).
BRIDGMAN, P. W., *The Nature of Physical Theory* (New York, 1936).
BROGLIE, L. DE, *Matter and Light* (New York, 1939).
BURNET, J., *Early Greek Philosophy* (London, 1948).
BURNET, J., *Greek Philosophy* (London, 1932).
BURTT, E. A., *The Metaphysical Foundations of Modern Physical Science* (London, 1949).
CARRÉ, M. H., *Realists and Nominalists* (Oxford, 1946).
CASSIRER, E., *Determinismus und Indeterminismus in der modernen Physik* (Göteborg, 1937).
CASSIRER, E., *An Essay on Man* (Yale, 1944).
CASSIRER, E., *The Problem of Knowledge* (Yale, 1950).
CHARLIER, C. L., *Die Mechanik des Himmels*, Band I (Leipzig, 1902).
COHEN and DRABKIN, *A Source Book in Greek Science* (New York, 1948).
COLLINGWOOD, R. G., *The Idea of Nature* (Oxford, 1945).
COMMINS and LINSCOTT (ed.), *The Philosophers of Science* (New York, 1947).
CORNFORD, F. M., *From Religion to Philosophy* (London, 1912).
CORNFORD, F. M., *Plato's Cosmology* (London, 1948).
DAMPIER, W. C., *A History of Science* (Cambridge, 1946).
DIRAC, P. A. M., *The Principles of Quantum Mechanics* (Oxford, 1947).
DUHEM, P., *La Théorie physique* (Paris, 1906).
FARRINGTON, B., *Science in Antiquity* (Oxford, 1947).
FRANK, P., *Modern Science and its Philosophy* (Harvard, 1949).
FREEMAN, K., *The Pre-Socratic Philosophers* (Oxford, 1946).
FRENKEL, J., *Wave Mechanics — Advanced General Theory* (Oxford, 1934).
GALILEO, *Dialogues concerning Two New Sciences*, transl. Crew and de Salvio (New York, 1933).
HEATH, T., *A History of Greek Mathematics*, vol. I (Oxford, 1921).
HEISENBERG, W., *Philosophic Problems of Nuclear Science* (London, 1952).
JEFFREYS, H. and B., *Methods of Mathematical Physics* (Cambridge, 1950).

JOHNSON, M., *Science and the Meanings of Truth* (London, 1946).

KNEDLER, J. W. (ed.), *Masterworks of Science* (Garden City, 1947).

LAUE, M. VON, *History of Physics* (New York, 1950).

LEIBNIZ, *Die Hauptwerke* (Leipzig, 1949).

LINDSAY and MARGENAU, *Foundations of Physics* (New York, 1936).

MACH, E., *The Analysis of Sensations* (Chicago, 1914).

MACH, E., *The Science of Mechanics* (La Salle, 1942).

MAGIE, W. F., *A Source Book in Physics* (New York, 1935).

MARGENAU, H., *The Nature of Physical Reality* (New York, 1950).

MISES, R. VON, *Positivism* (Harvard, 1951).

MOODY, E. A., *The Logic of William of Ockham* (London, 1935).

MOSER, S., *Grundbegriffe der Naturphilosophie bei Wilhelm von Occam* (Innsbruck, 1932).

PEIRCE, C. S., *Collected Papers*, vols. I, II, IV, V (Harvard, 1931–1934).

PLANCK, M., *General Mechanics*, vol. I of *Introduction to Theoretical Physics* (London, 1933).

PLANCK, M., *The Philosophy of Physics* (London, 1946).

PLANCK, M., *Scientific Autobiography and other Papers* (New York, 1949).

PLANCK, M., *The Universe in the Light of Modern Physics* (London, 1933).

PLANCK, M., *Vorträge und Erinnerungen* (Stuttgart, 1949).

PLANCK, M., *Where is Science going?* (London, 1933).

POINCARÉ, H., *Science and Hypothesis* (New York, 1905).

POINCARÉ, H., *Science and Method* (London, first publ. 1909).

RUARK and UREY, *Atoms, Molecules and Quanta* (New York, 1930).

RUSSELL, B., *A Critical Exposition of the Philosophy of Leibniz* (London, 1937).

RUSSELL, B., *A History of Western Philosophy* (London, 1946).

RUSSELL, B., *The Scientific Outlook* (London, 1949).

SCHLICK, M., *Allgemeine Erkenntnislehre* (Berlin, 1918).

SCHLICK, M., *Philosophy of Nature* (New York, 1949).

SCHRÖDINGER, E., *Abhandlungen zur Wellenmechanik* (Leipzig, 1928).

SCHRÖDINGER, E., *Science and Humanism* (Cambridge, 1951).

SIMON, M., *Geschichte der Mathematik im Altertum* (Berlin, 1909).

SOMMERFELD, A., *Atomic Structure and Spectral Lines* (London, 1934).

WENTZEL, G., *Quantum Theory of Fields* (New York, 1949).

WESTAWAY, F. W., *Scientific Method* (London, 1912).

WEYL, H., *Philosophy of Mathematics and Natural Science* (Princeton, 1949).

WEYL, H., *Space—Time—Matter* (New York, 1950).

WHITTAKER, E. T., *A Treatise on the Analytical Dynamics of Particles and Rigid Bodies* (New York, 1944).
WHITTAKER, E. T., *From Euclid to Eddington* (Cambridge, 1949).
WILSON, W., *Theoretical Physics*, vols. I–II (London, 1931).
ZELLER, E., *Outlines of the History of Greek Philosophy* (London, 1948).

Proof of the Euler-Lagrange Conditions for an Integral to be Stationary

OUR first object is to prove equation (55), which states that, if we are given two fixed points (x_1, t_1), (x_2, t_2) and a function $f\left(x, \dfrac{dx}{dt}, t\right)$, and we require to find a curve $x = x(t)$ through the fixed points such that the integral of the function f along the curve between the fixed points is stationary, the desired curve must satisfy the differential equation

$$\frac{d}{dt}\left(\frac{\partial f}{\partial\left(\dfrac{dx}{dt}\right)}\right) - \frac{\partial f}{\partial x} = 0. \qquad . \qquad . \qquad (55)$$

In our usual notation,

$$\delta\int_{t_1}^{t_2} f\left(x, \frac{dx}{dt}, t\right) dt = \int_{t_1}^{t_2}\left\{\frac{\partial f}{\partial x}\delta x + \frac{\partial f}{\partial\left(\dfrac{dx}{dt}\right)} \cdot \delta\left(\frac{dx}{dt}\right)\right\} dt$$

$$= \int_{t_1}^{t_2}\left\{\frac{\partial f}{\partial x}\delta x + \frac{\partial f}{\partial\left(\dfrac{dx}{dt}\right)}\frac{d}{dt}(\delta x)\right\} dt$$

$$= \int_{t_1}^{t_2}\left\{\frac{\partial f}{\partial x}\delta x - \frac{d}{dt}\left(\frac{\partial f}{\partial\left(\dfrac{dx}{dt}\right)}\right)\delta x + \frac{d}{dt}\left(\frac{\partial f}{\partial\left(\dfrac{dx}{dt}\right)}\delta x\right)\right\} dt$$

$$= \int_{t_1}^{t_2}\left\{\frac{\partial f}{\partial x} - \frac{d}{dt}\left(\frac{\partial f}{\partial\left(\dfrac{dx}{dt}\right)}\right)\right\}\delta x\, dt + \frac{\partial f}{\partial\left(\dfrac{dx}{dt}\right)}\delta x\, \bigg|_{t_1}^{t_2}.$$

The last term vanishes, δx being zero at t_1 and t_2, and, since δx may vary arbitrarily along the curve, we can write

$$\frac{d}{dt}\left(\frac{\partial f}{\partial\left(\frac{dx}{dt}\right)}\right) - \frac{\partial f}{\partial x} = 0$$

as a necessary and sufficient condition for the integral to be stationary.

When there are several dependent variables $x, y, \ldots$, and the integral $\int f\left(x, y, \ldots, \frac{dx}{dt}, \frac{dy}{dt}, \ldots, t\right) dt$, taken between two fixed points $(x_1, y_1, \ldots, t_1)$, $(x_2, y_2, \ldots, t_2)$ is to be stationary, we obtain, proceeding in exactly the same way,

$$\delta\int_{t_1}^{t_2} f\left(x, y, \ldots, \frac{dx}{dt}, \frac{dy}{dt}, \ldots, t\right) dt$$

$$= \sum_{x, y, \ldots} \int_{t_1}^{t_2}\left\{\frac{\partial f}{\partial x}\delta x + \frac{\partial f}{\partial\left(\frac{dx}{dt}\right)}\delta\left(\frac{dx}{dt}\right)\right\} dt$$

$$= \sum_{x, y, \ldots} \int_{t_1}^{t_2}\left\{\frac{\partial f}{\partial x}\delta x + \frac{\partial f}{\partial\left(\frac{dx}{dt}\right)}\frac{d}{dt}(\delta x)\right\} dt$$

$$= \sum_{x, y, \ldots} \int_{t_1}^{t_2}\left\{\frac{\partial f}{\partial x} - \frac{d}{dt}\left(\frac{\partial f}{\partial\left(\frac{dx}{dt}\right)}\right)\right\}\delta x\, dt$$

$$+ \sum_{x, y, \ldots}\frac{\partial f}{\partial\left(\frac{dx}{dt}\right)}\delta x\Big|_{t_1}^{t_2}.$$

Again, the last term vanishes, and, as δx, δy, $\ldots$ may vary arbitrarily and independently along the curve, the variational condition is equivalent to the equations

$$\left.\begin{aligned}\frac{d}{dt}\left(\frac{\partial f}{\partial\left(\frac{dx}{dt}\right)}\right) - \frac{\partial f}{\partial x} &= 0\\[2mm]\frac{d}{dt}\left(\frac{\partial f}{\partial\left(\frac{dy}{dt}\right)}\right) - \frac{\partial f}{\partial y} &= 0\\[2mm]\cdots\cdots\cdots\cdots\cdots\cdots\end{aligned}\right\}\qquad\cdot\quad\cdot\quad\cdot\quad(56)$$

In the third case, we have several independent variables $s, t, \ldots$, and wish to determine

$$x = x(s, t, \ldots)$$
$$y = y(s, t, \ldots)$$
$$\cdots \cdots \cdots,$$

so that the integral

$$\int f\left(x, y, \ldots, \frac{\partial x}{\partial s}, \frac{\partial x}{\partial t}, \ldots, \frac{\partial y}{\partial s}, \frac{\partial y}{\partial t}, \ldots, s, t, \ldots\right) ds\,dt \ldots,$$

taken over a fixed range of the independent variables, at whose boundaries the dependent variables are fixed, is stationary. The procedure is then analogous to that of the last problem. Thus

$$\delta \int f ds\,dt \ldots = \sum_{x, y, \ldots} \int \left\{ \frac{\partial f}{\partial x} \delta x + \sum_{s, t, \ldots} \frac{\partial f}{\partial\left(\dfrac{\partial x}{\partial s}\right)} \delta\left(\frac{\partial x}{\partial s}\right) \right\} ds\,dt \ldots$$

$$= \sum_{x, y, \ldots} \int \left\{ \frac{\partial f}{\partial x} \delta x + \sum_{s, t, \ldots} \frac{\partial f}{\partial\left(\dfrac{\partial x}{\partial s}\right)} \frac{\partial}{\partial s} (\delta x) \right\} ds\,dt \ldots$$

$$= \sum_{x, y, \ldots} \int \left\{ \frac{\partial f}{\partial x} - \sum_{s, t, \ldots} \frac{\partial}{\partial s} \left(\frac{\partial f}{\partial\left(\dfrac{\partial x}{\partial s}\right)} \right) \right\} \delta x\, ds\,dt \ldots$$

$$+ \sum_{x, y, \ldots} \int \left\{ \sum_{s, t, \ldots} \frac{\partial}{\partial s} \left(\frac{\partial f}{\partial\left(\dfrac{\partial x}{\partial s}\right)} \delta x \right) \right\} ds\,dt \ldots$$

As in the previous instances, the last term can easily be seen to vanish, because $\delta x, \delta y, \ldots$ are zero at the boundaries of the range of integration. The conditions for the integral to be stationary are therefore

$$\left.\begin{aligned}
\frac{\partial}{\partial s}\left(\frac{\partial f}{\partial\left(\dfrac{\partial x}{\partial s}\right)}\right) + \frac{\partial}{\partial t}\left(\frac{\partial f}{\partial\left(\dfrac{\partial x}{\partial t}\right)}\right) + \cdots - \frac{\partial f}{\partial x} = 0 \\[3mm]
\frac{\partial}{\partial s}\left(\frac{\partial f}{\partial\left(\dfrac{\partial y}{\partial s}\right)}\right) + \frac{\partial}{\partial t}\left(\frac{\partial f}{\partial\left(\dfrac{\partial y}{\partial t}\right)}\right) + \cdots - \frac{\partial f}{\partial y} = 0 \\[3mm]
\cdots \cdots \cdots \cdots \cdots \cdots \cdots
\end{aligned}\right\} \quad \cdots \ (57)$$

In conformity with the usual methods of applied mathematics, a rigorous treatment of these equations has not been given here, and differentials have been employed freely. We refer those who desire a more strict analysis of the subject, and also of the connexion between minimum and variational conditions, to mathematical discussions of the calculus of variations, such as can be found, for instance, in chapter VII, vol. II of R. Courant's *Differential and Integral Calculus*.

APPENDIX 2

Variational Principles and Chemical Reactions[1]

IT is the purpose of this investigation to determine the extent to which chemical processes can be subsumed under the category of variational principles. However, we do not propose to consider the detailed atomic rearrangement occurring in any particular reaction, and consequently the quantum-mechanical aspects of molecular structure are ignored. As thermodynamics does not depend on the atomic constitution of material systems, it probably provides the most apposite means for our inquiry into chemical reactions.

Now, the very terms which arise in chemical kinetics and thermodynamics, e.g., reaction co-ordinate, driving force and velocity of reaction already indicate an analogy between a dynamical system and a chemical process. Likewise, Hinshelwood[2] has pointed to the "real chemical inertia" which tends to slow down chemical reactions. And finally, from the work of Glasstone, Laidler and Eyring[3] it seems that a chemical process can be conceived as the movement of a point along the reaction co-ordinate, i.e., the reaction path drawn in one plane.

THERMODYNAMIC APPROACHES TO THE PROBLEM

Historically, the application of minimal principles to thermodynamics was first undertaken by Helmholtz[4] who showed for *reversible* processes that the Lagrangian function was analogous to the negative "Helmholtz free energy" F. He then suggested that all *irreversible* processes might be regarded as quasi-reversible; only the limitations of our technical tools prevent us from reversing the molecular process in a chemical reaction. In a different context, a somewhat similar idea is embodied in Tolman's principle[5] of microscopic reversibility.

In a recent paper, Fényes[6] has extended Helmholtz' rather vague minimal considerations to a variational principle

$$\delta \int F dt = 0.$$

This equation is equivalent to the statement that $\Delta F = 0$ (F a minimum), for reversible processes at constant temperature and volume. Further, he introduces a more general form of this variational condition for the case of an irreversible process, namely

$$\delta \int (F + f) dt = 0,$$

where f is a function which expresses the restriction of the movement of the mechanical system (to which thermodynamics is formally analogous) by two non-holonomic constraints. However, on closer scrutiny we observe that the above-mentioned equation —like the so-called "differential principles"—cannot be integrated, and therefore any pursuit of this approach is not likely to prove profitable. We shall presently return to the Helmholtz-Fényes formulation of a thermodynamic variational principle.

As an alternative attitude towards the thermodynamics of natural irreversible processes, the work of De Donder, Prigogine and others evades the exclusive employment of dynamics as the archetype of thermodynamics. Prigogine and Defay[7] define the extent of a chemical reaction at any moment t in terms of the "reaction co-ordinate" or "degree of advancement" ξ, by

$$\xi = \frac{n_i - n_i^0}{\nu_i},$$

where n_i is the number of moles of the ith component at any time t, n_i^0 the number of moles at time $t = 0$, and ν_i the stoichiometric coefficient. This coefficient is negative for reactants and positive for products. The advantage of using such a quantity is that it has the same value for each substance participating in the reaction.[8] The velocity of reaction is defined by $v = (d\xi/dt)$.

For a closed system, the differential of the entropy may be written

$$dS = d_e S + d_i S,$$

where $d_e S = dQ/T$ represents the entropy change due to the external heat change dQ, and $d_i S = dQ'/T$ the entropy change due to irreversible processes occurring within the reaction system

(dQ' is referred to as the "uncompensated heat"). Restricting ourselves to an isolated system, we have

$$dS = d_iS > 0.$$

If the only change that takes place in the system is due to chemical reaction, then the production of entropy is determined solely by $d\xi$. Accordingly,

$$dQ' = Td_iS = Ad\xi > 0.$$

In this relation a function of state A, called the affinity, is introduced by De Donder.[9] The affinity is defined by the relation

$$A = -\sum_i \nu_i\mu_i = -(\partial G/\partial \xi)_{p,T},$$

where μ_i is the chemical potential of the ith component and G is the Gibbs free energy. The inequality mentioned above is due to the fact that the entropy production is always positive for an irreversible process. Hence,

$$A(d\xi/dt) = Av > 0.$$

This De Donder inequality is the most significant relation in Prigogine's development of chemical thermodynamics.[7] The affinity A can be designated as the "driving force" of a chemical reaction. The product Av will, of course, be zero at equilibrium. In the form $Av \geqslant 0$, this relation is the basis of Prigogine's treatment of "theorems of moderation," of which the most outstanding is the principle of Le Chatelier-Braun. Although De Donder's inequality cannot be written as a variational principle, it has nevertheless been demonstrated by Prigogine and by Denbigh[10] that the product Av is a minimum for processes not too far removed from the equilibrium state.

In his latest monograph, Prigogine[11] deals tentatively with variational principles in irreversible thermodynamics. He introduces the additional property of the entropy production per unit time which we may express as $d_i^2S/dt^2 \leqslant 0$, and attempts to find a function L with the properties of the entropy production (d_iS/dt) mentioned above. This function is tantamount to a more general rate of entropy production. In a stable, stationary state, L, of course, is a minimum. For certain cases where L can be derived, it appears to be a function of the square of the rate of the irreversible process. Therefore, this square has a minimum value in the stationary state. Prigogine's speculation that this fact is reminiscent of Gauss' principle of least constraint is not very helpful for our purposes, as the Gauss theorem is, after all, a differential principle in no way related to a true variational principle.

We are now in a better position to evaluate Bak's derivation of a minimum principle for non-equilibrium steady states.[12] Using a one-dimensional diffusion equation, he establishes that such a state is accompanied by a minimum in a characteristic function which may be described as the sum of "resistance" times "fluxes" squared, $\sum_i R_i J_i^2$. This function corresponds to the entropy production, except for a constant factor, subject to the necessary condition that the reaction is near true equilibrium. Whence we may be tempted to conclude that in a chemical process we have a minimum principle for conditions *close enough* to equilibrium, as in the case of Hamilton's principle, where a real minimum is attained, provided the path of the system is *short enough*. It is worthy of note, however, that a steady state near equilibrium is hardly a general case of an irreversible process.

In a recent discussion, Prigogine[13] has dealt with non-linear problems in the thermodynamics of irreversible processes. He finds that stationary non-equilibrium states are characterized by a minimum of a quadratic function of the rates of irreversible processes. Near equilibrium, as in Bak's work, this function is simply the entropy production. In the most general case, the function is found to be

$$m = -\sum_i J_i \dot{X}_i,$$

where J_i is the rate and $\dot{X}_i$ is the time derivative of the generalized force or affinity. This function is formulated for the case of a single chemical reaction in the following section.

The Classical Mechanical Analogy

We now propose to develop a rigorous treatment of a chemical reaction as a classical mechanical process. Let us firstly define the velocity and acceleration of a chemical reaction by

$$v = d\xi/dt \qquad \text{and} \qquad \dot{v} = d^2\xi/dt^2,$$

respectively. Secondly, we assume that we may identify the Lagrangian function with the Gibbs free energy G. The justification of this assumption is provided by our subsequent work. This quantity G may be regarded, in general, as a function of the degree of advancement ξ, the velocity $d\xi/dt$, and the time t. We here postulate that conditions of constant temperature and pressure hold throughout this discussion. We may therefore express the variational principle in the form

$$\delta \int G(\xi, d\xi/dt, t)\, dt = 0.$$

This principle is equivalent to a set of differential equations (the Euler-Lagrange conditions):

$$\frac{d}{dt}\left(\frac{\partial G}{\partial(d\xi/dt)}\right) - \frac{\partial G}{\partial \xi} = 0.$$

Utilizing our previous definition of reaction velocity, the above equation becomes

$$\frac{d}{dt}\left(\frac{\partial G}{\partial v}\right) - \frac{\partial G}{\partial \xi} = 0,$$

which is simply "Lagrange's equation of motion for a chemical reaction". On this basis, we may extend our analogy by defining a "chemical momentum"

$$p_{\text{chem}} = (\partial G/\partial v).$$

In the more specialized, usual case, however, the Gibbs function depends only on ξ and is independent of time. It follows that the above Lagrange equation of motion reduces to

$$-(\partial G/\partial \xi) \equiv A = 0.$$

From Prigogine's paper referred to above,[13] it seems that, for a chemical reaction, the quantity characterized by a minimum in the stationary state is Av (equivalent to the entropy production) near equilibrium, or, in the general case, $-v\dot{A}$, a function which vanishes in the stationary state, and is positive in all non-stationary situations. This function has been discussed by the authors in connexion with the acceleration of a chemical reaction.[14] It appeared that the time rate of change of the affinity, $\dot{A}$, exists in the case of a general accelerated reaction.[15] Consequently, in such a general case, the Gibbs free energy is a function of time.

The method adopted here (which is a formal elaboration of the Helmholtz-Fényes approach in terms of modern thermodynamic concepts) leads any purely formal axiomatic treatment of the "congruency" between mechanics and thermodynamics to a logical conclusion. Yet, at the same time, it reveals the wide gulf between conceptual rigour and physical reality.

The foregoing remarks imply that it is possible to view a chemical reaction as the motion of a "point" along one dimension (viz., the reaction co-ordinate) in a "reaction space" with the co-ordinates x, y, ξ, where x and y are any two independent variables of state. It is essentially this concept which is employed in the field of absolute reaction rates.[3] There it is found that the reaction system follows "the most favourable reaction path on the potential energy surface"; in other words, this route—the reaction co-ordinate—signifies the path of minimal potential energy. Like trends of reasoning are encountered in all processes which involve minimum expenditure of some quantity. Such a conception is undoubtedly attractive to workers in chemical kinetics and has, moreover, proved fruitful in the treatment of rate processes. This transition state theory, with its so-called "saddle-point" and its reaction path of minimum energy, yields perhaps the nearest analogy to a true variational principle for a chemical reaction.

In spite of all these superficial resemblances, any rigid attempt at reducing chemical reactions *in toto* to dynamical systems and thus to variational principles is doomed to failure. Resorting to pictorial representation of a chemical process as a means of illustrating the macroscopic process is of value only as a broad auxiliary simplification facilitating the interrelation of theory and experimental data. Any deeper understanding of a chemical reaction demands recourse to quantum-mechanical methods which operate in the microscopic field of molecular structure.

It is thus only in a restricted sense that minimum principles, in the form of "least" theorems, can be derived for the equilibrium or steady state situation. Variational theorems—and this includes minimum principles—are no less and no more than means of summarizing existing results.

REFERENCES

1. YOURGRAU, W. and RAW, C. J. G., reprinted from *Nuovo Cimento*, **5** Supp. 3 (1957), 472.
2. HINSHELWOOD, C. N., *The Structure of Physical Chemistry* (Oxford, 1951).
3. GLASSTONE, LAIDLER and EYRING, *The Theory of Rate Processes* (New York, 1941).
4. HELMHOLTZ, *Crelle's Journal*, **100** (1886), 137.
5. TOLMAN, R., *The Principles of Statistical Mechanics* (Oxford, 1938).
6. FÉNYES, I., *Z. Phys.*, **132** (1952), 140.
7. PRIGOGINE, I. and DEFAY, R., *Chemical Thermodynamics* (London, 1954).
8. DENBIGH, K. G., *The Thermodynamics of the Steady State* (London, 1951).

9. DE DONDER, TH., *Bull. Acad. Roy. Belg. Cl. Sci.*, (5) **7** (1922), 197, 205.
10. DENBIGH, K. G., *Trans. Faraday Soc.*, **48** (1952), 389.
11. PRIGOGINE, I., *Introduction to Thermodynamics of Irreversible Processes* (Springfield, 1955).
12. BAK, T. A., *J. Phys. Chem.*, **59** (1955), 665.
13. PRIGOGINE, I., *Heat Transfer and Fluid Mechanics Institute Papers*, p. 261 (Stanford, 1956).
14. RAW, C. J. G. and YOURGRAU, W., *Nature*, **178** (1956), 809.
15. YOURGRAU, W. and RAW, C. J. G., *Nature*, **181** (1958), 480.

Index